අංක. ०५

MULTICRITERION OPTIMIZATION IN ENGINEERING
with FORTRAN programs

ELLIS HORWOOD SERIES IN MECHANICAL ENGINEERING

STRENGTH OF MATERIALS: Vol. 1 Fundamentals, Vol. 2 Applications
J. M. ALEXANDER, University College of Swansea.
TECHNOLOGY OF ENGINEERING MANUFACTURE
J. M. ALEXANDER, University College of Swansea, G. W. ROWE, Birmingham University and R. C. BREWER
VIBRATION ANALYSIS AND CONTROL SYSTEM DYNAMICS
C. BEARDS, Imperial College of Science and Technology
STRUCTURAL VIBRATION ANALYSIS
C. BEARDS, Imperial College of Science and Technology
COMPUTER AIDED DESIGN AND MANUFACTURE 2nd Edition
C. B. BESANT, Imperial College of Science and Technology
BASIC LUBRICATION THEORY 3rd Edition
A. CAMERON, Imperial College of Science and Technology
SOUND AND SOURCES OF SOUND
A. P. DOWLING and J. E. FFOWCS-WILLIAMS, University of Cambridge
MECHANICAL FOUNDATIONS OF ENGINEERING SCIENCE
H. G. EDMUNDS, University of Exeter
ADVANCED MECHANICS OF MATERIALS 2nd Edition
Sir HUGH FORD, F.R.S., Imperial College of Science and Technology, and J. M. ALEXANDER, University College of Swansea.
MECHANICAL FOUNDATIONS OF ENGINEERING SCIENCE
H. G. EDMUNDS, Professor of Engineering Science, University of Exeter
ELASTICITY AND PLASTICITY IN ENGINEERING
Sir HUGH FORD, F.R.S. and R. T. FENNER, Imperial College of Science and Technology
TECHNIQUES OF FINITE ELEMENTS
BRUCE M. IRONS, University of Calgary, and S. AHMAD, Bangladesh University, Dacca
FINITE ELEMENT PRIMER
BRUCE IRONS and N. SHRIVE, University of Calgary
CONTROL OF FLUID POWER: ANALYSIS AND DESIGN 2nd (Revised) Edition
D. McCLOY, Ulster Polytechnic, N. Ireland and H. R. MARTIN, University of Waterloo, Ontario, Canada
UNSTEADY FLUID FLOW
R. PARKER, University College, Swansea
DYNAMICS OF MECHANICAL SYSTEMS 2nd Edition
J. M. PRENTIS, University of Cambridge
ENERGY METHODS IN VIBRATION ANALYSIS
T. H. RICHARDS, University of Aston Birmingham
ENERGY METHODS IN STRESS ANALYSIS: With Intro. to Finite Element Techniques
T. H. RICHARDS, University of Aston in Birmingham
COMPUTATIONAL METHODS IN STRUCTURAL AND CONTINUUM MECHANICS
C. T. F. ROSS, Portsmouth Polytechnic
ENGINEERING DESIGN FOR PERFORMANCE
K. SHERWIN, Liverpool University
ROBOTS AND TELECHIRS
M. W. THRING, Queen Mary College, University of London

MULTICRITERION OPTIMIZATION IN ENGINEERING
with FORTRAN programs

ANDREZJ OSYCZKA, M.Sc., Ph.D.
Associate Professor
Department of Mechanical Engineering
University of Cracow, Warsaw

Translation Editor:
B.J. DAVIES
Professor of Mechanical Engineering
University of Manchester Institute of Science and Technology

ELLIS HORWOOD LIMITED
Publishers · Chichester

Halsted Press: a division of
JOHN WILEY & SONS
New York · Chichester · Brisbane · Toronto

First published in 1984 by
ELLIS HORWOOD LIMITED
Market Cross House, Cooper Street, Chichester, West Sussex, PO19 1EB, England

The publisher's colophon is reproduced from James Gillison's drawing of the ancient Market Cross, Chichester.

Distributors:

Australia, New Zealand, South-east Asia:
Jacaranda-Wiley Ltd., Jacaranda Press,
JOHN WILEY & SONS INC.,
G.P.O. Box 859, Brisbane, Queensland 40001, Australia

Canada:
JOHN WILEY & SONS CANADA LIMITED
22 Worcester Road, Rexdale, Ontario, Canada.

Europe, Africa:
JOHN WILEY & SONS LIMITED
Baffins Lane, Chichester, West Sussex, England.

North and South America and the rest of the world:
Halsted Press: a division of
JOHN WILEY & SONS
605 Third Avenue, New York, N.Y. 10016, U.S.A.

© 1984 A. Osyczka/Ellis Horwood Limited

British Library Cataloguing in Publication Data
Osyczka, Andrzej
Multicriterion optimization in engineering with FORTRAN programs. –
(Ellis Horwood series in mechanical engineering)
1. Engineering systems – Mathematical models
2. FORTRAN (Computer program language)
I. Title
620.7'2'0285424 TA168
Library of Congress Card No. 83-25478

ISBN 0-85312-481-7 (Ellis Horwood Limited)
ISBN 0-470-20019-7 (Halsted Press)

Printed in Great Britain by Butler & Tanner, Frome, Somerset

Contents

Contents

Preface

Multicriterion optimization problems arise in different engineering fields and scientists are beginning to devote considerable attention to developing methods for solving them. The aim of these methods is to help the engineer to make the right decision in conflicting situations, i.e., in situations in which several objectives must be satisfied. Among many different decision-making problems the class of mathematical programming problems has recently begun to receive much attention. In this class of problems an optimization task is described by functions which refer both to constraints and objectives and which give a formal description of the task. Such problems will be discussed in this book.

As in many fields of operational research, the theory of multicriterion optimization is more developed and better represented in the literature than is the case of engineering applications. Moreover there is a large number of different methods, and this book does not claim to cover them all, but it is oriented towards the basic methods and their engineering applications.

This book is intended to serve as an introductory text to the topic of multicriterion optimization. It is written for a person interested in operational research but without an extensive mathematical background. The presentation of material should be easy for engineers to understand and many simple examples are given.

Chapter 1 is an introduction to multicriterion optimization problems for those who are new to the subject. The basic terms used in multicriterion nomenclature such as decision variables, constraints and objective functions, etc., are explained to help the reader to describe an engineering problem in a formalized way.

This formalized description is the subject of Chapter 2. Concepts of the Pareto and min-max optimum are introduced in order to determine which solutions are worth seeking.

Methods discussed in Chapters 3 and 4 provide the engineer with such solutions. These methods are arbitrarily divided into two groups based on (i) function scalarization and (ii) min-max approach. Numerical methods only are considered here since analytical methods for large problems prove unsatisfactory.

The formalized description discussed in Chapter 2 does not cover all the optimization problems the reader may meet. For a very large group of engineering problems network modelling reflects their nature best. Such problems are discussed in Chapter 5 in which the basic concept of multicriterion network optimization is presented and a method of seeking the Pareto and min-max paths is provided.

Chapter 6 presents four authentic problems modelled by means of the multicriterion approach and solved by using some of the methods discussed in the book.

Finally in the appendices three Fortran programs for the methods presented in Chapters 4 and 5 are described and listed. All the programs have been tested and run on the CDC Cyber 72 computer at CYFRONET – Cracow and on the CDC 7600 and Cyber 72 computers at UMRCC – Manchester.

Chapters 4, 5 and 6 are based on the research done by the author and his associates.

It is possible that some errors have been made; the author will be grateful if they are brought to his notice.

The preparation of this book was a lengthy process and several people influenced the finished product. I am deeply indebted to Professor B. J. Davies for his encouragement, and the help in preparing the final version of the manuscript.

I thank Mr J. Zając for his help when testing the computer programs. There are many other colleagues from my university whose help and discussion have been invaluable in preparing this book.

I am also grateful to the many students who through the years have laboured over some examples and computer programs included in the book.

I would like to thank Professor D. M. Himmelblau for his generous permission to implement his computer program as a part of the interactive computer system presented in Appendix B.

Some financial support for my investigations came from the British Council in the form of modest but greatly appreciated grants, and it is my pleasure to thank that organization.

Finally I would like to apologize to my wife and children for depriving them of many evenings and weekends while preparing the manuscript.

CHAPTER 1

Introduction to the formulation of the multicriterion optimization problem

In complex engineering optimization problems there often exist several non-commensurable criteria which must be considered. This situation is formulated as a multicriterion optimization problem (also called multi-performance, multiple objective or vector optimization) in which the engineer's goal is to minimize and/or maximize not a single objective function but several functions simultaneously.

Formulation of an optimization problem consists in constructing a mathematical model that describes the behaviour of a physical system encompassing the problem area. This model must closely approximate the actual behaviour of the system for the solution obtained to be adequate and useful.

In this chapter it will be shown how to formulate the multicriterion optimization problem for mathematical programming; in other words how to build the mathematical model of the system. Generally speaking, mathematical programming in the sense considered here is the analysis of problems of the type: 'Find the optimum of objective functions when the decision variables are subject to inequality and equality constraints.' Mathematical programming is rapidly becoming both a practicable and fundamental tool in engineering studies and other fields.

Throughout this chapter three simple examples are continued in order to help the reader understand the language of mathematical programming. These examples are strictly related to real engineering problems, however, for the sake of convenience they have been considerably simplified and thus they do not cover the scope of complexity of engineering problems which can be solved by the methods described in following chapters.

1.1 DECISION VARIABLES

In an engineering optimization task the numerical quantities for which values are to be chosen will be called decision variables or more shortly variables. In mathematical programming these quantities are denoted as $x_i, i = 1, 2, ..., n$, Thus x_i is a variable representing the ith quantity.

1.1.1 Examples

To illustrate the concept of the decision variables consider the following three examples. The first is a production planning problem to determine how many units to produce of each of two products, denoted as A and B. We could then define

x_1 = number of units of product A to produce,
x_2 = number of units of product B to produce.

The second example is a beam design problem to determine dimensions of the beam shown in Fig. 1.1 where

x_1 = length of part 1 of the beam,
x_2 = interior diameter of the beam.

The third is a metal cutting problem to determine the cutting conditions for a lathe in which a single cutting tool turns a diameter in one pass. The decision variables are

x_1 = cutting speed,
x_2 = feed per revolution.

1.1.2 Decision variables and parameters

In some optimization models the choice of the number and type of decision variables is simple but quite often the situation is not very clear. Consider

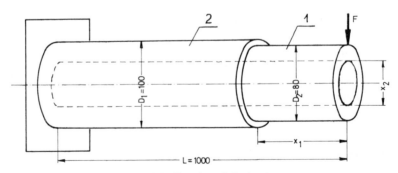

Fig. 1.1 Drawing of the beam

the second example in which the exterior diameter of the beam are assumed to be 100 mm and 80 mm. If these diameters are not predetermined, i.e., can be chosen in the process of design, then they can be treated as the third and fourth decision variables. A similar question arises when we consider the beam material. If the designer is free to choose the material of the beam from the given set of materials then we shall have a new decision variable.

At the stage of building the optimization model we have to decide which quantities are treated as decision variables and which are taken as fixed. The quantities whose values are fixed will be called parameters. Mathematical relations between the decision variables and the parameters constitute an engineering optimization model.

Quantities may be designated as parameters for various reasons. A common one is that we are simply not at liberty to change that particular quantity. In some cases it may be known from experience that a particular value of the quantity always gives good results, hence, there is no reason to treat this quantity as the decision variable. Sometimes, at the beginning of building the optimization model, it is difficult to decide which quantities are to be treated as the parameters and which as decision variables. Generally speaking, if we are free to choose (within a range) the values of any quantity then this quantity should be treated as a decision variable. However, for the simplicity's sake we can assume that some quantities will be treated as parameters even if they could be considered as decision variables. In this way the optimization model is easier to solve but the results may not be strictly related to reality. Consider the third example in which we have two decision variables. The third one could be the depth of the cut. But then the optimization model will be much more complicated and it is not always necessary to use such a model. It is for the engineer to decide what degree of simplification is allowed.

1.1.3 *Formal notation*

Further we shall also refer to a vector of decision variables $\bar{x}$, which is simply a column containing all the variables in a particular problem. For the first example the $\bar{x}$ might be defined as

$$\bar{x} = \begin{bmatrix} x_1 \\ x_2 \end{bmatrix}$$

The order of the components of the vector x is arbitrary, but once the make-up of the vector is defined any specific vector may be said to be a 'solution'. For instance

$$\bar{x} = \begin{bmatrix} 50 \\ 150 \end{bmatrix}$$

specifies for the first example a particular production programme. When we have n variables then

$$\bar{x} = \begin{bmatrix} x_1 \\ x_2 \\ \cdot \\ \cdot \\ \cdot \\ x_n \end{bmatrix}$$

or written in a more convenient way $\bar{x} = [x_1, x_2, ..., x_n]^T$, where T denotes the transporsition of the column vector to the row vector.

1.2 CONSTRAINTS

In each engineering task there are some restrictions dictated by environment, processes and/or resources which must be satisfied in order to produce an acceptable solution. These restrictions are collectively called constraint functions or more shortly constraints, and describe dependences among decision variables and parameters. These dependences are written in the form of mathematical inequalities and sometimes also equalities.

1.2.1 Examples

Consider the production planning problem. Both products A and B require time in two departments. Product A requires 1 hour in the first department and $1\frac{1}{4}$ hours in the second department. Product B requires 1 hour in the first department and $\frac{3}{4}$ hour in the second department. The available hours in each department are 200 monthly. Furthermore, there is a maximum market potential of 150 units for product B. These restrictions can be written as the inequality constraints as follows

$$
\begin{aligned}
x_1 + x_2 &\leqslant 200 \\
1\tfrac{1}{4}x_1 + \tfrac{3}{4}x_2 &\leqslant 200 \\
x_2 &\leqslant 150 \\
x_1 &\geqslant 0 \\
x_2 &\geqslant 0
\end{aligned}
\tag{1.1}
$$

In the second example it is assumed that the beam in Fig. 1.1 should resist the maximum force $F_{max} = 12\,000$ N and the permissible bending stress of the beam material is $\sigma_g = 180$ N/mm. Thus, the bending strength constraints are:

for part 1

$$\frac{F_{max}x_1}{\dfrac{D_2^4 - x_2^4}{32D_2}} \leqslant \sigma_g \qquad (1.2)$$

and for part 2

$$\frac{F_{max}l}{\dfrac{D_1^4 - x_2^4}{32D_1}} \leqslant \sigma_g \qquad (1.3)$$

After substitution we have

$$\frac{9.78 \cdot 10^6 x_1}{4.096 \cdot 10^7 - x_2^4} \leqslant 180 \qquad (1.4)$$

$$x_2 \leqslant 75.2$$

Assuming also that the interior diameter of the beam is to be no less than 40 mm and that the length of part 1 can be freely chosen we have the geometric constraints

$$x_2 \geqslant 40 \qquad (1.5)$$
$$x_1 \geqslant 0$$

For the metal cutting problem the constraints imposed on x_1 and x_2 by the machine tool are as follows:

(i) Assuming that the maximum and the minimum available cutting speed are 263.9 and 21.1 m/min respectively we have

$$x_1 \leqslant 263.9 \qquad (1.6)$$
$$x_1 \geqslant 21.1$$

(ii) Similarly the constraints imposed on the feed are

$$x_2 \leqslant 1.0$$
$$x_2 \geqslant 0.05 \tag{1.7}$$

where it is assumed that the maximum and the minimum available feeds are 1.0 and 0.05 mm/rev.

(iii) We also have the power constraint. The power consumed in cutting can be determined from the following equation

$$P = \frac{F_t x_1}{60\,000} \text{ kW} \tag{1.8}$$

where F_t is the tangential cutting force which can be determined as

$$F_t = C x_2^{\alpha} d^{\beta} \text{ N} \tag{1.9}$$

where C, α and β are constants, d is the depth of the cut which is held constant at a given value. The denominator in equation (1.8) is the product 60×1000 when speeds are given in m/min and forces in N.

If P_{max} is the maximum power available at the spindle of the machine tool, then the power constraint is

$$x_1 x_2^{\alpha} \leqslant \frac{60\,000 P_{max}}{C d^{\beta}} \tag{1.10}$$

Assuming that $P_{max} = 8$ kW, $d = 6$ mm, $C = 1000$, $\alpha = 0.75$ and $\beta = 0.9$ we have

$$x_1 x_2^{0.75} \leqslant 95.7 \tag{1.11}$$

The process requirement imposes on x_1 and x_2 a constraint which ensures the stable sutting region. Iwata et al. (1979) suggested that the constraint of the form

$$x_1^{\delta} x_2 \geqslant \gamma \tag{1.12}$$

should be used to avoid those x_1 and x_2 values that are likely to cause chatter vibration, adhesion and build-up-edge formulation. δ and γ are constants to be estimated.

Assuming that $\delta = 2.0$ and $\gamma = 1600$, the above constraint is

$$x_1^2 x_2 \geqslant 1600 \tag{1.13}$$

We may also have workpiece requirement constraints, for example the required surface roughness and geometric accuracy of the workpiece. If the model concerns rough machining usually no such constraints will be imposed on x_1 and x_2.

The general form of writing the inequality constraints is

$$g_j(\bar{x}) \geqslant 0 \qquad \text{for } j = 1, ..., m \tag{1.14}$$

Thus, the inequalities given by (1.1) will have their general form as follows[†]

$$g_1(\bar{x}) \equiv 200 - x_1 - x_2 \geqslant 0$$
$$g_2(\bar{x}) \equiv 200 - 1\tfrac{1}{4}x_1 - \tfrac{3}{4}x_2 \geqslant 0$$
$$g_3(\bar{x}) \equiv 150 - x_2 \geqslant 0 \tag{1.15}$$
$$g_4(\bar{x}) \equiv x_1 \geqslant 0$$
$$g_5(x) \equiv x_2 \geqslant 0$$

1.2.2 Equality constraints

In certain models we can also have the equality constraints, which will be written in the following form

$$h_j(\bar{x}) = 0 \qquad j = 1, ..., p \tag{1.16}$$

Consider the second example. We have inequality constraints

$$g_1(\bar{x}) \equiv 180 - \frac{9.78 \times 10^6 x_1}{4.096 \times 10^7 - x_2^4} \geqslant 0$$
$$g_2(\bar{x}) \equiv 75.2 - x_2 \geqslant 0 \tag{1.17}$$
$$g_3(\bar{x}) \equiv x_2 - 40 \geqslant 0$$
$$g_4(\bar{x}) \equiv x_1 \geqslant 0$$

Furthermore, the designer can be faced with an additional restriction that the length of part 1 must be five times greater than the interior diameter of the beam. Then, we have the equality constraint

$$h_1(\bar{x}) \equiv x_1 - 5x_2 = 0 \tag{1.18}$$

[†]The symbol $\equiv$ is intended to be read 'is defined as'.

Note that p, the number of equality constraints, must be less than n, the number of decision variables. If $p \geq n$ the problem is termed as over-constrained, since there are no degrees of freedom left for optimizing. The number of degrees of freedom is given by $n - p$.

1.2.3 *Implicit form of constraints*

In the examples considered all the constraints are written in explicit forms. There are, however, more complicated models for which it is not possible to write the explicit dependence of g_j on the components of vector $\bar{x}$. Still for many numerical methods of optimization, implicit forms of constraints are satisfactory. Obviously, these forms should allow the calculation of $g_j(\bar{x})$ for any given vector $\bar{x}$ and this is usually done by means of an algorithm.

Consider the third example. If we want to build the model in which the shape error of the workpiece is considered as a new constraint then this constraint will have an implicit form. Assuming only the theoretical shaft displacement, caused by normal cutting force F_n, the diameter variations designated by ΔD may be expressed by the equation (see Kaczmarek (1976))

$$\Delta D = 2F_n\left[c_h\left(\frac{y}{l}\right)^2 + c_t\left(\frac{l-y}{l}\right)^2 + c_s + \frac{1}{3EI}\frac{(l-y)^2}{l}y^2\right] \qquad (1.19)$$

where: c_h = headstock compliance,

c_t = tailstock compliance,

c_s = saddle compliance,

l = length of the workpiece,

E = Young's modulus,

I = moment of the inertia of the workpiece,

y = actual load position.

The saddle compliance is constant and independent of the load position, i.e., independent of y. The tailstock and headstock compliances can be assumed to be constant, and changes of displacement occur only as a result of force variations acting on both assemblies as load position varies. Headstock and tailstock load changes may be approximated by reactions of a beam supported at two points and acted on by a moving force. If we assume a constant cutting force, when turning a uniform shaft, the increment of diameter as compared with the value established at the starting point, i.e., at the tailstock centre, will vary (see Fig. 1.2).

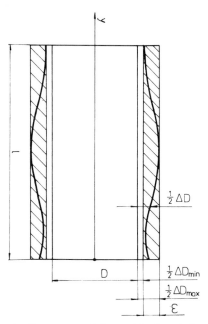

Fig. 1.2 Shape error of the machined workpiece

The equation for the normal cutting force F_n is

$$F_n = C_1 x_2^\alpha d^\beta \qquad (1.20)$$

where C_1, α_1 and β_1 are constants. Assuming that $c_1 = 500$, $\alpha_1 = 0.7$, $\beta_1 = 0.85$ and $d = 6$ mm we have

$$F_n = 2293 x_2^{0.7} \, \text{N} \qquad (1.21)$$

Furthermore, if we assume that for the machine tool the compliance values in N/mm are $c_h = 2.0 \times 10^{-5}$, $c_t = 2.5 \times 10^{-5}$, $c_s = 1.5 \times 10^{-5}$ and that the diameter and the length of the workpiece are 60 and 500 mm respectively, then after substitution to equation (1.19) we have

$$\Delta D = 4586 x_2^{0.7} \left[2.0 \times 10^{-5} \left(\frac{y}{500} \right)^2 + 2.5 \times 10^{-5} \left(\frac{500 - y}{500} \right)^2 \right.$$
$$\left. + 1.5 \times 10^{-5} + 2.55 \times 10^{-12} \frac{(500 - y)^2}{500} y^2 \right] \text{mm} \qquad (1.22)$$

The shape error ϵ is given by

$$\epsilon = \Delta D_{max} - \Delta D_{min} \qquad (1.23)$$

where ΔD_{min} and ΔD_{max} denote the minimum and the maximum value of diameter increment respectively.

If we want this error to be no greater than 0.05 mm, then we have the constraint

$$\Delta D_{max} - \Delta D_{min} \leqslant 0.05 \qquad (1.24)$$

This constrraint is in the implicit form because the value of D depend not only on x_2 but also on y which is not a decision variable. Hence, it is necessary to write an algorithm which will determine the values of ΔD_{min} and of ΔD_{max} for any given value of x_2. The simplest algorithm is as follows:

(1) Do Steps 2, 3 and 4, for $y = l,\ l - \Delta y,\ l - 2\Delta y, ..., 0$ where Δy is an arbitrarily chosen step size.
(2) Calculate ΔD using formula (1.22).
(3) If $\Delta D < \Delta D_{min}$, then $\Delta D_{min} = \Delta D$.
(4) If $\Delta D > \Delta D_{max}$, then $\Delta D_{max} = \Delta D$.

1.3 OBJECTIVE FUNCTIONS

In the process of selecting a 'good solution' from all solutions which satisfy the constraints, there must be some criteria which allow these solutions to be compared. These criteria are the inherent qualities of each solution and in the optimization model thay must be expressed as computable functions of the decision variables. These functions, called the objective functions, are apparently non-commensurable and usually some of them will be in conflict with others. We designate the objective functions as $f_1, f_2, ..., f_k$ or to emphasize their dependence upon the decision variables, as $f_1(\bar{x}), f_2(\bar{x}), ..., f_k(\bar{x})$.

1.3.1 *Examples*

Consider the production planning example in which product A is of high quality and product B is of lower quality. The profits are $4 and $5 per product respectively. The best customer of the company wishes to have as many as possible of type A product. We realize that two objectives: (1) the maximization of profit, and (2) the maximum production of product A, should be considered in this problem. Hence, the objective functions can be written as follows

$$f_1(\bar{x}) = 4x_1 + 5x_2$$
$$f_2(\bar{x}) = x_1 \qquad\qquad (1.25)$$

In the beam design example the designer realizes that the beam should satisfy two objectives:

(1) the minimization of the volume of the beam, and
(2) the minimization of the static compliance of the beam.

The function describing the volume of the beam is

$$f_1(\bar{x}) = \frac{\pi}{4} [x_1(D_2^2 - x_2^2) + (l - x_1)(D_1^2 - x_2^2)] \qquad (1.26)$$

The static compliance of the beam for the displacement under the force F is

$$f_2(\bar{x}) = \frac{64}{3\pi E} \left[\left(\frac{1}{D_2^4 - x_2^4} - \frac{1}{D_1^4 - x_2^4} \right) x_1^3 + \frac{l^3}{D_1^4 - x_2^4} \right] \qquad (1.27)$$

where E is Young's modulus.

We assumed that $l = 1000\,\mathrm{mm}$, $D_1 = 100\,\mathrm{mm}$, $D_2 = 80\,\mathrm{mm}$ and $E = 2.06 \cdot 10^5\,\mathrm{N/mm^2}$ and thus we obtain

$$f_1(\bar{x}) = 0.785 [x_1(6400 - x_2^2) + (1000 - x_1)(10\,000 - x_2^2)]\,\mathrm{mm^3} \quad (1.28)$$

$$f_2(\bar{x}) = 3.298 \times 10^{-5} \left[\left(\frac{1}{4.096 \times 10^7 - x_2^4} - \frac{1}{10^8 - x_2^4} \right) x_1^3 + \frac{10^9}{10^8 - x_2^4} \right] \mathrm{mm/N}$$
$$(1.29)$$

For the metal cutting problem we can assume two objectives: (1) maximization of metal removal rate and (2) maximization of tool life. The first objective function is

$$f_1(\bar{x}) = 1000 d x_1 x_2 \, \mathrm{mm^3/min} \qquad (1.30)$$

Since we assume that depth of cut d is equal $6\,\mathrm{mm}$ we have

$$f_1(\bar{x}) = 6000 x_1 x_2 \qquad (1.31)$$

The tool life equation is usually written as follows

$$f_2(\bar{x}) = \frac{A}{x_1^{n_1} x_2^{n_2}} \min \tag{1.32}$$

where A, n_1 and n_2 are constants.

Assuming that $A = 1.28 \times 10^7$, $n_1 = 3.33$ and $n_2 = 2.22$ the second objective function is

$$f_2(\bar{x}) = \frac{1.28 \times 10^7}{x_1^{3.33} x_2^{2.22}} \tag{1.33}$$

1.3.2 Objective and constraint functions

Since single criterion optimization methods are highly developed, many engineering problems are reduced to single criterion optimization models which require the selection of only one quality as the objective function, which can often be quite difficult, whereas in multicriterion optimization models this difficulty does not arise as we can assume as many objective functions as we wish. The only question which may occur is which qualities should be treated as objective functions and which as constraints. Consider the last example in which the number of workpieces produced between tool changes may be assumed as the third objective function. This function can be expressed as follows

$$f_3(\bar{x}) = \frac{A\lambda}{l x_1^{n_1 - 1} x_2^{n_2 - 1}} \text{ workpieces} \tag{1.34}$$

where: A, n_1 and n_2 as in equation (1.32),

$\lambda = 1000/(\pi D)$ where D is the mean workpiece diameter (mm),
l = distance travelled by the tool in making a turning pass (mm).

Assuming that $l = 500$ and $D = 60\,\text{mm}$ we have

$$f_3(\bar{x}) = \frac{1.358 \times 10^5}{x_1^{2.33} x_2^{1.22}} \tag{1.35}$$

In case we want to restrict the number of workpieces produced between tool changes saying, for example, that there should not be less than 20 workpieces, then the function (1.35) will be introduced to the model as a new constraint of the form

$$\frac{1.358 \times 10^5}{x_1^{2.33} x_2^{1.22}} \geqslant 20 \qquad (1.36)$$

Furthermore, for some machining processes this function can be neglected in the model as inessential. These three cases will form three different models for which optimal solutions will be different. It is up to the engineer to decide which model more closely approximates real conditions.

Note that we can also consider another model in which the shape error of the workpiece is a new objective function. In this case the objective function will be in an implicit form. Numerical methods discussed in this book allow both constraint and objective functions to be in this form.

1.3.3 *Formal notation*

The objective functions form a vector function $\bar{f}(\bar{x})$ which is a column containing all the functions considered in the model. As for the vector of decision variables $\bar{x}$, the order of the components of the vector $\bar{f}(\bar{x})$ is arbitrary, but once we choose this order we are not allowed to change it. Of course, this choice is made while prescribing the index to each particular objective function. Hence, for the first example the vector function is

$$\bar{f}(\bar{x}) = \begin{bmatrix} f_1(\bar{x}) \\ f_2(\bar{x}) \end{bmatrix}$$

where: $f_1(\bar{x}) = $ profit,
$\qquad f_2(\bar{x}) = $ number of product A.

Generally, for the kth objective functions we have

$$\bar{f}(\bar{x}) = \begin{bmatrix} f_1(\bar{x}) \\ f_2(\bar{x}) \\ \vdots \\ f_k(\bar{x}) \end{bmatrix}$$

or written in a more convenient way $\bar{f}(\bar{x}) = [f_1(\bar{x}), f_2(\bar{x}), ..., f_k(\bar{x})]^T$.

1.4 SPACE OF DECISION VARIABLES AND SPACE OF OBJECTIVE FUNCTIONS

The set of all n-tuples of real numbers denoted by E^n is called Euclidean n-space. Since we have the vector of the decision variables and the vector of the objective functions we have to consider two Euclidean spaces: (1) the n-dimensional space of the decision variables in which each coordinate axis corresponds to a component of vector $\bar{x}$, and (2) the k-dimensional space of the objective functions in which each coordinate axis corresponds to a component of vector $\bar{f}(\bar{x})$. Every point in the first space represents a solution and gives a certain point in the second space which determines a quality of this solution in terms of the values of the objective functions.

1.4.1 *Examples*

Since in all our examples both spaces are two-dimensional we can have graphical illustrations of their problem formulation. For the first example the space of decision variables is shown in Fig. 1.3(a). In this space the inequality constraints given by (1.15) form a set of feasible solutions, in other words, a feasible region denoted by X and designated by the shaded area. Any point $\bar{x}$ in this region $\bar{x}\epsilon X$ defines a feasible solution, i.e., the solution which satisfies all constraints. If for this point we calculate values of the objective functions using formulas given by (1.25), these values will give us a point in the space of objective functions Fig. 1.3(b). All the points from the set of feasible solutions X, form in the space of objective functions, a set of their feasible values denoted by F and designated in Fig. 1.3(b) by the shaded area.

Similarly the graphical illustration of the problem formulation is shown in Fig. 1.4 for the beam design problem for which the constraints are given by (1.17), and the objective functions are given by (1.28) and (1.29).

If we now consider another model of the same example in which we have additionally the equality constraint given by (1.18), then the graphical illustration of the problem formulation will be different (Fig. 1.5). In this case the set of feasible solutions X is limited to the points on the heavy line in Fig. 1.5(a). In the space of objective functions (Fig. 1.5(b)) only the curve represents the set of their feasible values. We can easily see how equality constraints limit the set of feasible solutions.

Finally, consider the metal cutting problem. The inequality constraints given by (1.6), (1.7), (1.11) and (1.13) form the set of feasible solutions X as presented in Fig. 1.6(a). This set forms set F as presented in Fig. 1.6(b) for the objective functions given by (1.31) and (1.33).

Multicriterion optimization for mathematical programming in simple cases, i.e., for two-or three-dimensional spaces, can have a graphical

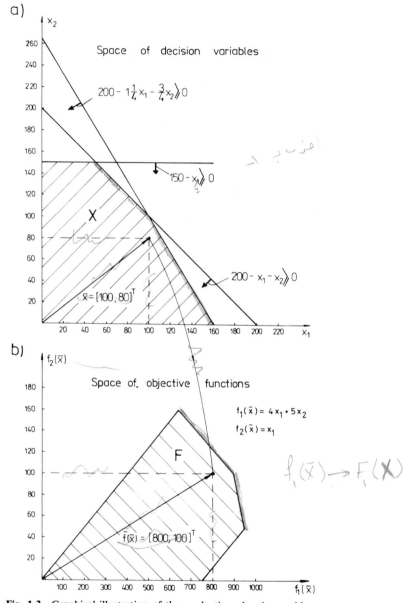

Fig. 1.3 Graphical illustration of the production planning problem

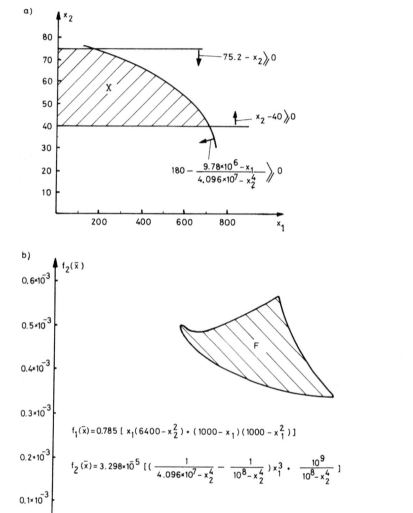

Fig. 1.4 Graphical illustration of the beam design problem

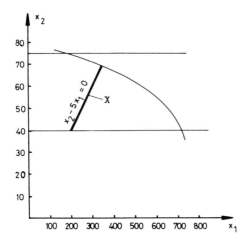

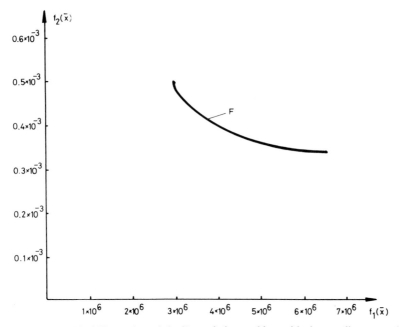

Fig. 1.5 Graphical illustration of the beam design problem with the equality constraint

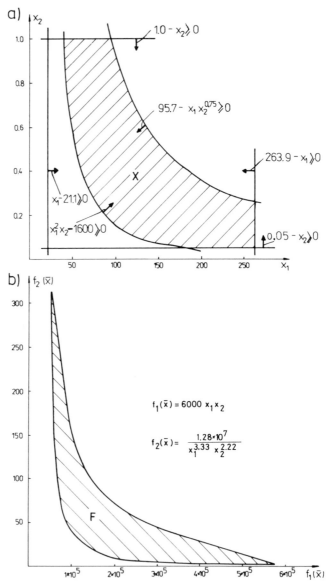

Fig. 1.6 Graphical illustration of the metal cutting problem

illustration of the problem formulation. This illustration will also help us to understand the methods of solving multicriterion optimization problems discussed in following chapters. In more complicated cases, when spaces more than three-dimensional are considered, we can interpret the problem formulation in the same way but it is not possible to have a graphical illustration of it.

1.5 THE INTEGRATED PROBLEM FORMULATION

After making the appropriate engineering judgements and defining all necessary objective functions and constraint functions we formulate a multicriterion optimization problem for mathematical programming as follows.

Find the vector $\bar{x}^* = [x_1^*, x_2^*, ..., x_n^*]^\mathrm{T}$ which will satisfy the m inequality constraints:

$$g_j(\bar{x}) \geqslant 0 \qquad j = 1, 2, ..., m \qquad (1.37)$$

the p equality constraints

$$h_j(\bar{x}) = 0 \qquad j = 1, 2, ..., pn \qquad (1.38)$$

and optimize the vector function

$$\bar{f}(\bar{x}) = [f_1(\bar{x}), \qquad f_2(\bar{x}), ..., f_k(\bar{x})]^\mathrm{T}$$

where $\bar{x} = [x_1, x_2, ..., x_n]^\mathrm{T}$ is the vector of decision variables.

In other words we wish to determine from among the set of all numbers which satisfy (1.37) and (1.38) that particular set $x_1^*, x_2^*, ..., x_k^*$ which yields the optimum values of all the objective functions.

Here 'optimize' does not mean simply to find the minimum or maximum of the objective function as it is for a single criterion optimization problem. It means to find a 'good' solution considering all the objective functions. Of course, first we need to know how to designate a particular solution as 'good' or 'bad'.

This is the major question which arises while solving any multicriterion optimization problem, and will be discussed in the next chapter.

Assuming that we already know what the optimum is, the next question is to find this optimim or a set of optimal solutions which helps the engineer to make a decision. This leads to the methods of problem solving which will be discussed in Chapters 3 and 4.

In the standard problem formulation given above both inequality and equality constraints appear. Of course, we shall also deal with problem in which we have (1) only inequality constraints, (2) only equality constraints and (3) no constraints. In the third case, called an unconstrained optimization problem, there are no restrictions imposed on vector $\bar{x}$ and then the set of feasible solutions X is all of Euclidean space E^n. In most optimization problems of physical significance such a case occurs very rarely. Note that when the upper and the lower values of a decision variable are assumed then we have to deal with inequality constraints.

1.5.1 *Forms of mathematical programming*

The problem formulation presented here refers to mathematical programming which can generally be classified as follows:

(1) Linear programming where both the objective and constraint functions are linear functions of decision variables, see the first example.
(2) Non-linear programming where at least one of the functions mentioned above is non-linear, see the second and third examples. Note that in the third example the constraint and objective functions can be linearized by taking logarithms, and in this case this example can be classified as belonging to linear programming.
(3) Integer programming where it is required that the solution be chosen not from the set of real numbers but from the set of integers. Numerical methods for solving such problems are more complicated and even for single criterion optimization problems are much less highly developed than those for linear and non-linear programming.
(4) Discrete programming where it is required that the solution be chosen from a permitted set of discrete values. This programming is the most complex to solve both for single and multicriterion optimization problems, however, it appears quite often in engineering tasks.

1.5.2 *Example*

Consider the metal cutting problem. In the case in which only discrete values of spindle speeds and feeds can be set on the lathe on which the workpiece is machined this problem must be considered as discrete programming. Assuming, for example, that on the lathe the spindle speeds can be chosen from the set

{112, 140, 180, 224, 280, 355, 450, 560, 710, 900, 1120, 1400} rev/min

and multiplying the elements of this set by $\pi D/1000$ we obtain the following

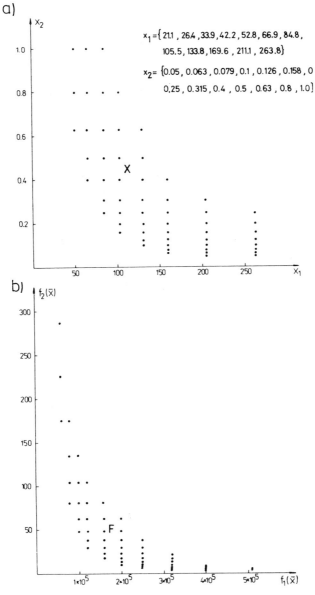

Fig. 1.7 The metal cutting problem considered as a discrete programming problem

set of discrete values of the first decision variable

$$x_1 = \{21.1, \ 26.4, \ 33.9, \ 42.2, \ 52.8, \ 66.9, \ 84.8,$$
$$105.5, \ 133.8, \ 169.6, \ 211.1, \ 263.8\} \ \text{m/min}$$

In a similar manner the lathe may have the following set of feeds set on

$$x_2 = [0.05, \ 0.063, \ 0.079, \ 0.1, \ 0.126, \ 0.158, \ 0.2, \ 0.25, \ 0.315,$$
$$0.4, \ 0.5, \ 0.63, \ 0.8, \ 1.0] \ \text{mm/rev}$$

Then the metal cutting problem considered as discrete programming can be illustrated graphically as shown in Fig. 1.7, where the dots represent feasible solutions.

If for integer or discrete programming some of the variables are allowed to be real-valued, then we have to deal with mixed integer or mixed discrete programming.

For all these cases the multicriterion decision-making problem is the same, but the methods of dealing with each case are different and vary in the level of difficulty.

CHAPTER 2

Multicriterion mathematical programming problem

In this chapter the multicriterion mathematical programming problem is formulated as follows: find a vector of decision variables which satisfies constraints and optimizes a vector function whose elements represent the objective functions. These functions form a mathematical description of performance criteria which are usually in conflict with each other. Hence, the term 'optimize' means finding such a solution which would give acceptable values for all the objective functions. We shall discuss what a mathematically formalized approach can offer here. First, the Pareto optimum is defined. This optimum gives a set of non-inferior solutions, i.e., solutions for which there is no way of improving any criterion without worsening at least one other criterion. It is clear that the solution should be chosen from this set but the choice is still great. Secondly, the min-max optimum is defined. This optimum gives one solution which treats all the criteria as equally important. For this solution the values of all the objective functions are as close as possible to their separately attainable minima. If the solution is not acceptable, which happens fairly often, the engineer is then faced with the decision-making problem which is discussed at the end of this chapter.

2.1 PROBLEM FORMULATION

A multicriterion optimization problem for mathematical programming can be formulated as follows:

Find $\bar{x}^*$ such that

$$\bar{f}(\bar{x}^*) = \text{opt } \bar{f}(\bar{x}) \tag{2.1}$$

and such that

$$g_j(\bar{x}) \geqslant 0 \qquad j = 1, 2, \ldots, m \tag{2.2}$$

$$h_j(\bar{x}) = 0 \qquad j = 1, 2, \ldots, p < n \tag{2.3}$$

where $\bar{x} = [x_1, x_2, \ldots, x_n]^T$ is a vector of decision variables defined in n-dimensional Euclidean space of variables E^n, $\bar{f}(\bar{x}) = [f_1(\bar{x}), \ldots, f_i(\bar{x}), \ldots, f_k(\bar{x})]^T$ is a vector function defined in k-dimensional Euclidean space of objectives E^k, $g_j(\bar{x})$, $h_j(\bar{x})$ and $f_i(\bar{x})$ are linear and/or non-linear functions of variables $x_1, x_2, \ldots, x_n$.

The constraints given by (2.2) and (2.3) define the feasible region X and any point $\bar{x}$ in X defines a feasible solution. The vector function $\bar{f}(\bar{x})$ is a function which maps the set X in the set F which represents all possible values of the objective functions. The k components of the vector $\bar{f}(\bar{x})$ represent the non-commensurable criteria which must be considered. The constraints $g_j(\bar{x})$ and $h_j(\bar{x})$ represent the restriction imposed on the decision variables.

We reserve $\bar{x}^*$ to denote the optimal solution (or set of optimal solutions, since the solution may not be unique). We also use $I = [1, 2, \ldots, k]$ to denote the set of indices for all the objective functions.

In multicriterion optimization problems we may have to (1) minimize all the objective functions, (2) maximize all the objective functions and (3) minimize some and maximize others. For the sake of convenience we shall convert all the functions which are to be maximized into the form which

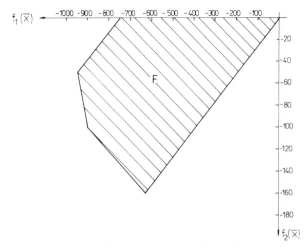

Fig. 2.1 Set F for the production planning problem after employing (2.4)

allows their minimization. This can be done by employing the identity

$$\max f_i(\bar{x}) = -\min(-f_i(\bar{x}))$$ (2.4)

For example, applying (2.4) to the originally stated problem as presented in Fig. 1.3 we obtain the set F as shown in Fig. 2.1.

Similarly, the inequality constraints of the form

$$g_j(\bar{x}) \leqslant 0 \qquad j = 1, 2, \ldots, m$$ (2.5)

can be converted to (2.2) form by multiplying by -1 and changing the sign of the inequality. Thus (2.5) is equivalent to

$$-g_j(\bar{x}) \geqslant 0 \qquad j = 1, 2, \ldots, m$$ (2.6)

In fact, the p equality constraints given by (2.3) could be replaced by the $2p$ inequalities

$$h_j(\bar{x}) \geqslant 0 \qquad j = 1, 2, \ldots, p$$
$$-h_j(\bar{x}) \leqslant 0 \qquad j = 1, 2, \ldots, p$$ (2.7)

Since numerical methods for dealing with equality and inequality constraints differ, we prefer to consider these two types of constraints separately.

Many optimization methods can cope only with inequality constraints. These methods are less complicated than those dealing also with equality constraints. Models with only inequality constraints occur very often in engineering tasks.

A multicriterion optimization problem can be written in a shortened form as follows

$$\bar{f}(\bar{x}^*) = \operatorname*{opt}_{x \in X} \bar{f}(\bar{x})$$ (2.8)

where

$$\bar{f} : X \longrightarrow E^k$$
$$X = \{\bar{x} \in E^n \mid \bar{g}(\bar{x}) \geqslant 0, \ \bar{h}(\bar{x}) = 0\}$$

The abbreviation 'opt' means here, the optimum of the vector function. Conflicting situations result in there being no general definition of this optimum which can determine a universal goal to be aimed at for solving all multicriterion models. Of course, if there exists $\bar{x}^* \in X$ such that for all

$i = 1, 2, \ldots, k$

$$\bigwedge_{x \in X} (f_i(\bar{x}^*) \leqslant f_i(\bar{x})) \qquad (2.9)$$

then, $\bar{x}^*$ is certainly a desirable solution.

An example of this situation is illustrated in Fig. 2.2. Unfortunately, this is an utopian situation which rarely exists, because it is unlikely that all $f_i(\bar{x})$ will achieve their minimum values in X at a common point x^*. Thus we are almost always faced with the dilemma: What solution should we adopt; that is, how should an 'optimal' solution be defined?

Before we try to answer this question, let us explain two terms we shall refer to in this book.

First consider the so-called ideal solution. In order to determine this solution we have to find separately attainable minima, for all the objective functions. Assuming that these minima can be found, let $\bar{x}^{0(i)} = [x_1^{0(i)}, x_2^{0(i)}, \ldots, x_n^{0(i)}]^\mathrm{T}$ be a vector of variables which minimizes the ith objective function $f_i(x)$. In other words, the vector $\bar{x}^{0(i)} \in X$ is such that

$$f_i(\bar{x}^{0(i)}) = \min_{x \in X} f_i(\bar{x}) \qquad (2.10)$$

We use f_i^0 to denote the minimum value of the ith function.

A vector $\bar{f}^0 = [f_1^0, f_2^0, \ldots, f_k^0]^\mathrm{T}$ is ideal for a multicriterion optimization problem and the point in E^n which determined this vector is the ideal solution.

In order to find the vector $\bar{f}^0$ we have to solve k scalar optimization problems.

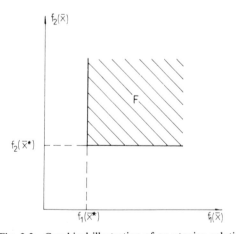

Fig. 2.2   Graphical illustration of an utopian solution

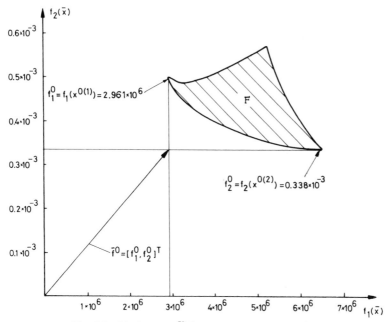

Fig. 2.3 Ideal vector $\bar{f}^0$ for the beam design problem

Since the ideal solution is not feasible we are not interested in finding it, but we often refer to the vector $\bar{f}^0$. Fig. 2.3 gives the graphical illustration of the vector f^0 for the problem presented in Fig. 1.4. The second term is convexity. The set F is convex if for every $\bar{u}^1$, $\bar{u}^2 \in F$ and every $\theta \in [0, 1]$

$$\bar{f}(\theta \bar{u}^1 + (1 - \theta)\bar{u}^2) \leqslant \theta \bar{f}(\bar{u}^1) + (1 - \theta)\bar{f}(\bar{u}^2) \tag{2.11}$$

In other words, the set F is convex if for any points $\bar{u}^1$ and $\bar{u}^2$ in the set, the line segment joining these points is also in the set. For example, the sets shown in Fig. 2.4(a) and (b) are convex. Those in Fig. 2.4(c) and (d) are not.

For the multicriterion optimization methods we have to know if the so-called t-directional shadow for the range F is convex. We denote this shadow by F^t and define it as follows:

$$F^t = \{ y \in E^k \,|\, y = \bar{f} + \alpha \bar{t}, \bar{f} \in F^p, \alpha \in E^1, \alpha \geqslant 0 \} \tag{2.12}$$

Fig. 2.5 illustrates graphically this definition.

We shall call the multicriterion optimization problem convex if F^t is convex for all $\bar{t} > 0$.

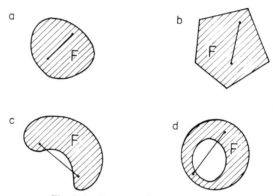

Fig. 2.4   Convex and non-convex sets

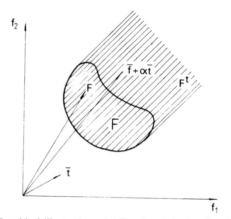

Fig. 2.5   Graphical illustration of t-directional shadow for the range F

Some methods of solution require the problem to be convex. Using these methods for non-convex problems no solution can be obtained. The dilemma is that there is no analytical method to classify a problem as being convex or non-convex.

2.2 PARETO OPTIMUM

The concept of this optimum was formulated by V. Pareto in 1896 and it is still the most important part of multicriterion analysis.

A common way of stating this optimum is contained in:

A point $\bar{x}^* \in X$ is Pareto optimal if for every $\bar{x} \in X$ either,

$$\bigwedge_{i \in I} (f_i(\bar{x}) = f_i(\bar{x}^*)) \tag{2.13}$$

or, there is at least one $i \in I$ such that

$$f_i(\bar{x}) > f_i(\bar{x}^*) \tag{2.14}$$

This definition is based upon the intuitive conviction that the point $\bar{x}^*$ is chosen as the optimal if no criterion can be improved without worsening at least one other criterion.

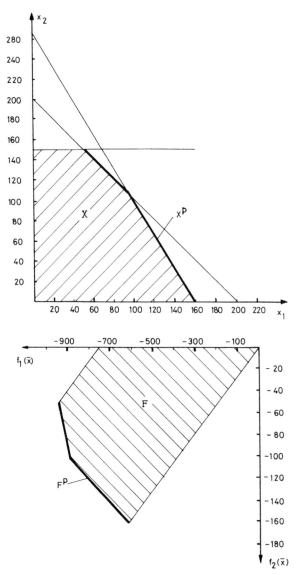

Fig. 2.6 The sets X^p and F^p for the production planning problem

Unfortunately the Pareto optimum almost always gives not a single solution but a set of solutions called non-inferior or non-dominated solutions. We use X^p to denote this set of solutions and F^p to denote the map of X^p in the space of objectives. Of course, the set X^p is derived from the set F^p which satisfies (2.13) and (2.14).

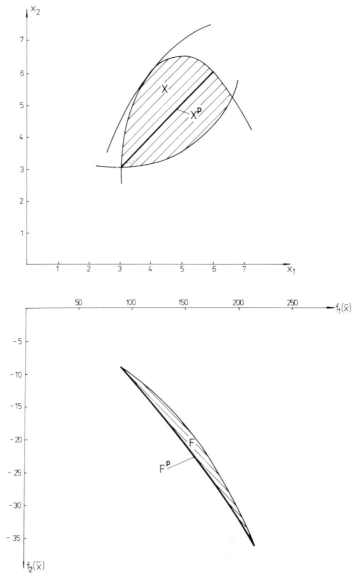

Fig. 2.7   The sets X^p and F^p for Example 2.1

Consider the problem presented in Fig. 1.3 for which in Fig. 2.6 both sets X^p and F^p are designated by heavy lines. Note that (2.13) and (2.14) are satisfied only when F^p lies on the boundary of F.

It often happens that X^p also lies on the boundary of X, but it is not a rule. Consider the following example.

Example 2.1

The problem is

Optimize

$$f_1(\bar{x}) = x_1^2 + x_2^2 + 12(x_1 + x_2) \longrightarrow \text{min}$$

$$f_2(\bar{x}) = x_1 x_2 \longrightarrow \text{max.}$$

Subject to

$$g_1(\bar{x}) \equiv -0.5x_1^2 + 5x_1 - x_2 - 6 \geqslant 0,$$

$$g_2(\bar{x}) \equiv -x_1^2 + 6x_1 - x_2^2 + 14x_2 - 42 \geqslant 0,$$

$$g_3(\bar{x}) \equiv -x_1^2 + 16x_1 - x_2^2 + 6x_2 - 48 \geqslant 0.$$

We transform the second objective function so that it could be minimized, i.e., we multiply it by -1. Hence we have

$$f_2(\bar{x}) = -x_1 x_2$$

The graphical illustration of the above problem is shown in Fig. 2.7, where the sets X^p and F^p are designated by heavy lines.

On the basis of the Pareto optimum important theorems in multicriterion optimization theory have been developed. Since their application to numerical methods is limited we shall not introduce them in this book.

2.3 MIN-MAX OPTIMUM

The idea of stating the min-max optimum and applying it to multicriterion optimization problems, was taken from game theory which deals with solving conflicting situations. The min-max approach to a linear model was proposed by Jutler (1967) and Solich (1969). It has been further developed by Osyczka (1978), (1981).

The min-max optimum compares relative deviations from the separately attainable minima. Consider the ith objective function for which the relative deviation can be calculated from

$$z_i'(\bar{x}) = \frac{|f_i(\bar{x}) - f_i^0|}{|f_i^0|}$$ (2.15)

or from

$$z_i''(\bar{x}) = \frac{|f_i(\bar{x}) - f_i^0|}{|f_i(\bar{x})|}$$ (2.16)

It is clear that for (2.15) and (2.16) we have to assume that for every $i \in I$ and for every $\bar{x} \in X$, $f_i(\bar{x}) \neq 0$.

Note that we transform the originally stated problem to the form in which all the objective functions are minimized. This transformation causes equations (2.15) and (2.16) to give different representations of the relative deviations. For the functions which are to be minimized equation (2.15) defines function relative increments, whereas, for those to be maximized it defines their relative decrements. Equation (2.16) works conversely.

Let $\bar{z}(\bar{x}) = [z_1(\bar{x}), ..., z_i(\bar{x}), ..., z_k(\bar{x})]^T$ to be a vector of the relative increments which are defined in E^k. The components of the vector $z(\bar{x})$ will be evaluated from the formula

$$\bigwedge_{i \in I} (z_i(\bar{x}) = \max\{z_i'(\bar{x}), z_i''(\bar{x})\}$$ (2.17)

Now we define the min-max optimum as follows.

A point $\bar{x}^* \in X$ is min-max optimal, if for every $\bar{x} \in X$ the following recurrence formula is satisfied:

Step 1

$$v_1(\bar{x}^*) = \min_{x \in X} \max_{i \in I} \{z_i(\bar{x})\}$$

and then $I_1 = \{i_1\}$, where i_1 is the index for which the value of $z_i(\bar{x})$ is maximal.

If there is a set of solutions $X_1 \subset X$ which satisfies Step 1, then

Step 2

$$v_2(\bar{x}^*) = \min_{x \in X_1} \max_{\substack{i \in I \\ i \notin I_1}} \{z_i(\bar{x})\}$$

and then $I_2 = \{i_1, i_2\}$, where i_2 is the index for which the value of $z_i(x)$ in this step is maximal .... (2.18)

If there is a set of solutions $X_{r-1} \subset X$ which satisfies step $r-1$ then

Step r

$$v_r(\bar{x}^*) = \min_{\substack{x \in X_{r-1} \\ i \notin I_{r-1}}} \max_{i \in I} \{z_i(\bar{x})\}$$

and then $I_r = \{I_{r-1}, i_r\}$, where i_r is the index for which the value of $z_i(\bar{x})$ in the rth step is maximal ....

If there is a set of solution $X_{k-1} \subset X$ which satisfies Step $k-1$, then

Step k

$$v_k(\bar{x}^*) = \min_{\bar{x} \in X_{k-1}} z_i(\bar{x}) \qquad \text{for } i \in I \text{ and } i \notin I_{k-1}$$

where $v_1(\bar{x}^*), ..., v_k(\bar{x}^*)$ is the set of optimal values of fractional deviations ordered non-increasingly.

This optimum can be described as follows. Knowing the extremes of the objective functions which can be obtained by solving the optimization problems for each criterion separately, the desirable solution is the one which gives the smallest values of the relative increments of all the objective functions.

The point $\bar{x}^* \in X$ which satisfies (2.18) may be called the best compromise solution considering all the criteria simultaneously and on equal terms of importance.

Formula (2.18) seems to be very complicated but in many applications it can be simplified (see Sections 3.4 and 4.2).

Example 2.2

Consider the following problem

Optimize

$$f_1(\bar{x}) = x_1 + x_2^2 \longrightarrow \min$$
$$f_2(\bar{x}) = x_1^2 + x_2 \longrightarrow \min$$

subject to

$$g_1(\bar{x}) \equiv 12 - x_1 - x_2 \geqslant 0$$
$$g_2(\bar{x}) \equiv -x_1^2 + 10x_1 - x_2^2 + 16x_2 - 80 \geqslant 0$$

To operate on a finite set of solutions we assume that this is an integer programming problem.

Table 2.1 Results of calculations for Example 2.2

No.	$\bar{x} = [x_1, x_2]^T$	$\bar{f}(\bar{x}) = [f_1(\bar{x}), f_2(\bar{x})]^T$	$z_1^i(x)$	$z_1^n(x)$	$z_2^i(x)$	$z_2^n(x)$	$\bar{z}(\bar{x}) = [z_1(\bar{x}), z_2(x)]^T$	$\max_{l \in \{1,2\}} \{z_i(x)\}$	i_1	$z_i(x), i \neq i_1$
1	$[2, 8]^T$	$[66.0, 12.0]^T$	1.200	0.545	0.000	0.000	$[1.200, 0.000]^T$	1.200	1	0.000
2	$[3, 6]^T$	$[39.0, 15.0]^T$	0.300	0.230	0.250	0.200	$[0.300, 0.250]^T$	0.300	1	0.250
3	$[3, 7]^T$	$[52.0, 16.0]^T$	0.733	0.423	0.333	0.250	$[0.733, 0.333]^T$	0.733	1	0.333
4	$[3, 8]^T$	$[67.0, 17.0]^T$	1.233	0.552	0.416	0.294	$[1.233, 0.416]^T$	1.233	1	0.416
5	$[3, 9]^T$	$[84.0, 18.0]^T$	1.800	0.642	0.500	0.333	$[1.800, 0.500]^T$	1.800	1	0.500
6	$[4, 6]^T$	$[40.0, 22.0]^T$	0.333	0.250	0.833	0.454	$[0.333, 0.833]^T$	0.833	2	0.333
7	$[4, 7]^T$	$[53.0, 23.0]^T$	0.766	0.433	0.916	0.478	$[0.766, 0.916]^T$	0.916	2	0.766
8	$[4, 8]^T$	$[68.0, 24.0]^T$	1.266	0.558	1.000	0.5	$[1.266, 1.000]^T$	1.266	1	1.000
9	$[5, 5]^T$	$[30.0, 30.0]^T$	0.000	0.000	1.500	0.6	$[0.000, 1.500]^T$	1.500	2	0.000
10	$[5, 6]^T$	$[41.0, 31.0]^T$	0.366	0.268	1.583	0.612	$[0.366, 1.583]^T$	1.583	2	0.366
11	$[5, 7]^T$	$[54.0, 32.0]^T$	0.800	0.444	1.666	0.625	$[0.800, 1.666]^T$	1.666	2	0.800
12	$[6, 6]^T$	$[42.0, 42.0]^T$	0.400	0.285	2.500	0.714	$[0.400, 2.500]^T$	2.500	2	0.400

For the above problem the separately attainable minima are

$$f_1^0 = 30 \qquad \bar{x}^{0(1)} = [5, 5]^T$$

$$f_2^0 = 12 \qquad \bar{x}^{0(2)} = [2, 8]^T$$

For all the integer values of $\bar{x}$ which satisfy constraints $g_1(\bar{x})$ and $g_2(\bar{x})$ Table 2.1 gives the results of calculations of the components of the vector $z(x)$. Now we choose all the solutions which satisfy the first step of the recurrence formula (2.18).

In this example we have only one solution, which is $\bar{x}^* = [3, 6]^T$. Hence, we do not have to refer to the second step of (2.18). For x^* the values of the objective functions are $f_1(x^*) = 39$ and $f_2(x^*) = 15$, and the values of the relative increments are $z_1(\bar{x}^*) = 0.300$ and $z_2(\bar{x}^*) = 0.250$.

The graphical illustration of this example is presented in Fig. 2.8.

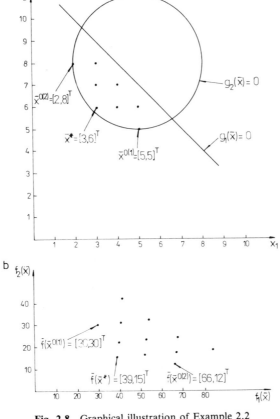

Fig. 2.8 Graphical illustration of Example 2.2

Table 2.2 Results of calculations for Example 2.1 considered as an integer programming problem

No.	$\bar{x} = [x_1, x_2]^T$	$\bar{f}(\bar{x}) = [f_1(\bar{x}), f_2(\bar{x})]^T$	$z_1^t(x)$	$z_1^t(\bar{x})$	$z_2^t(\bar{x})$	$z_2^{tt}(\bar{x})$	$\bar{z}(\bar{x}) = [z_1(x), z_2(\bar{x})]^T$	$\max\limits_{l \in \{1,2\}} \{z_l(x)\}$	i_1	$z_i(\bar{x}), i \neq i_1$
1	$[3, 3]^T$	$[\ 90.0,\ -9.0]^T$	0.000	0.000	0.75	3.000	$[0.000, 3.000]^T$	0.300	2	0.000
2	$[4, 4]^T$	$[128.0,\ -16.0]^T$	0.422	0.96	0.555	1.250	$[0.422, 1.250]^T$	1.250	2	0.422
3	$[4, 5]^T$	$[149.0,\ -20.0]^T$	0.655	0.395	0.444	0.800	$[0.655, 0.800]^T$	0.800	2	0.655
4	$[4, 6]^T$	$[172.0,\ -24.0]^T$	0.911	0.476	0.333	0.500	$[0.911, 0.500]^T$	0.911	1	0.500
5	$[5, 4]^T$	$[149.0,\ -20.0]^T$	0.655	0.395	0.444	0.800	$[0.655, 0.800]^T$	0.800	2	0.655
6	$[5, 5]^T$	$[170.0,\ -25.0]^T$	0.888	0.470	0.305	0.440	$[0.888, 0.400]^T$	0.888	1	0.400
7	$[5, 6]^T$	$[193.0,\ -30.0]^T$	1.144	0.533	0.166	0.200	$[1.144, 0.200]^T$	1.144	1	0.200
8	$[6, 5]^T$	$[193.0,\ -30.0]^T$	1.144	0.533	0.166	0.200	$[1.144, 0.200]^T$	1.144	1	0.200
9	$[6, 6]^T$	$[216.0,\ -36.0]^T$	1.400	0.583	0.000	0.000	$[1.400, 0.000]^T$	1.400	1	0.000

We can easily see that in this example for every $i \in I$ we have

$$z_i'(\bar{x}) > z_i''(\bar{x}) \tag{2.19}$$

This is the case in which for the originally stated problem all the objective functions are to be minimized. Thus formula (2.17) can be simplified to the following form

$$\bigwedge_{i \in I} (z_i(\bar{x}) = z_i'(\bar{x})) \tag{2.20}$$

Similarly if for the originally stated problem all the objective functions are to be maximized, we will have

$$z_i'(x) < z_i''(x) \tag{2.21}$$

for every $i \in I$ and thus formula (2.17) can be written as follows

$$\bigwedge_{i \in I} (z_i(\bar{x}) = z_i''(x)) \tag{2.22}$$

If the originally stated problem is a mixture of minimized and maximized functions formula (2.20) or (2.22) causes the vector $\bar{z}(\bar{x})$ to be a mixture of the function relative increments and decrements.

Consider Example 2.1 as an integer programming problem. Table 2.2 gives the results of calculations of the elements of $\bar{x}(\bar{x})$. We see that only for formula (2.17) will the elements of $\bar{z}(\bar{x})$ represent the relative increments for both functions.

Note that using (2.17) both the relative increments and decrements of all the objective functions are as small as possible.

In most optimization models the min−max optimum is determined in the first step of the formula (2.18). The case for a two criterion optimization is illustrated in Fig. 2.9(a). The models in which we have to refer to the second and/or the remaining steps of (2.18) are fairly rare (see Fig. 2.9(b)). Finally we can have the models in which after considering all the steps of (2.18) we still have more than one solution (see Fig. 2.9(c)). These solutions are equivalent considering the min−max optimum. Such models are very rare.

Note that the situations as presented in Fig. 2.9(b) and (c) can only happen for non-convex problems and will be more frequent for integer and discrete programming problems.

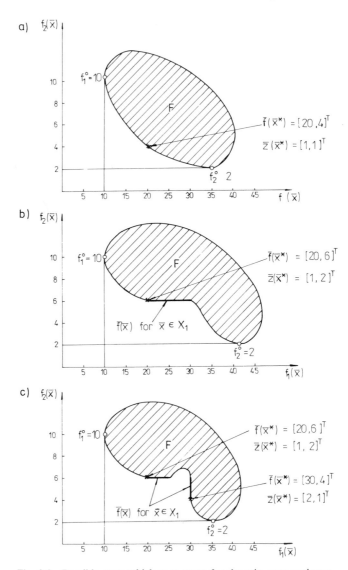

Fig. 2.9 Possible cases which may occur for the min−max optimum

Consider the problem illustrated in Fig. 1.7 in which we have two solutions which satisfy the first step of (2.18). These solutions are

$$\bar{x}^1 = [84.8, 0.315]^T \quad \text{for which } \bar{f}(\bar{x}^1) = [1.6 \times 10^5, 63.0]^T$$
$$\text{and } \bar{z}(\bar{x}^1) = [2.18, 2.54]^T,$$
$$\bar{x}^2 = [66.9, 0.4]^T \quad \text{for which } \bar{f}(\bar{x}^2) = [1.6 \times 10^5, 81.6]^T$$
$$\text{and } \bar{z}(\bar{x}^2) = [2.18, 1.72]^T.$$

The second step of (2.18) selects the solution $\bar{x}^2$ as the min–max optimum.

In Example 2.1 treated as an integer programming problem we have two solutions $\bar{x}^1 = [4, 5]^T$ and $\bar{x}^2 = [5, 4]^T$ which satisfy both steps of (2.18) (see Table 2.2). Note that the situation presented in Example 2.1 differs from the one shown in Fig. 2.9(c). In Example 2.1 both min–max optimal solutions give the same point in the space of objectives, whereas in Fig. 2.9(c) both min–max optimal solutions give different points in the space of objectives.

2.4 DECISION MAKING PROBLEM

After building a multicriterion optimization model of a physical system the engineer is faced with two questions: (1) which optimization technique to apply to solve this model, and (2) whether the solution obtained is satisfactory. For solving multicriterion optimization models single criterion optimization techniques are usually used. Since the literature on this subject is vast, the question of how to choose the appropriate technique to solve the model will not be discussed in this book. The reader may refer to the following textbooks: Dixon (1972), Fox (1971), Himmelblau (1972), Wismer and Chattergy (1978). The second question, which implies the acceptance of a solution, is called the decision-making problem.

In Sections 2.2 and 2.3 we have shown what the optima in the Pareto and min–max sense mean. It is clear that these optima will not give us a universal answer to the decision-making problem for all multicriterion optimization models. If all the criteria are equally important, then the min–max optima may define a desirable solution. In all other cases, a solution from the set of Pareto optimal solutions, i.e., from the set X^p, should be chosen. The complete set X^p has usually a very large number of solutions and it is impossible to find and consider all of them. It may be easier to find the desirable solution in the set X^p if the preferences as to the importance of criteria are known *a priori*. The information about these preferences may be required for some methods of solution. Since this information is usually incomplete and cannot be expressed in a fully formalized way, it rarely happens that a solution from the set X^p which we obtain using one of these methods, will be acceptable to the engineer. Thus most methods concentrate on providing a representative subset of X^p. Exploring this subset the engineer chooses a most preferred solution which he is prepared to admit as satisfactory.

For problems for which the information about preferences is given only *a posteriori* the first step is to obtain some solutions from the set X^p which guide the decision maker in further exploration.

Usually he would like to know the separately attainable minima and also the min–max optimum. Quite often the subset of X^p which covers the set F^p uniformly may be helpful in making the right decision. In some cases the decision maker may want to have more alternatives in the neighbourhood of a solution which seems to be close to the desirable one.

Note that the objective functions represent different physical qualities and thus it is usually difficult to make a judgement while comparing their values. It may be helpful for the decision-maker to know additionally the vector $\bar{z}(\bar{x})$ for each solution he explores, since non-dimensional values of components of $\bar{z}(\bar{x})$ make the comparison of solutions more convenient.

2.4.1 *Examples*

We shall now illustrate the decision-making problem using the examples from the previous chapter. Consider the metal cutting problem presented in Fig. 1.6. The engineer realizes that both criteria should be treated on terms of equal importance and thus he decides to seek the optimum in the min–max sense. This optimum gives the results

(1) Values of the decision variables

cutting speed $x_1^* = 57.9\,\text{m/min}$

feed $x_2^* = 0.476\,\text{mm/rev}$

(2) Values of the objective functions

metal removal rate $f_1(\bar{x}^*) = 1.65 \times 10^5\,\text{mm}^3/\text{min}$

tool life $f_2(\bar{x}^*) = 89.3\,\text{min}$

If in his opinion this solution is satisfying, there is no need for looking through other non-inferior solutions.

Consider the production planning problem presented in Fig. 1.3. The manager decides *a priori* that in this problem the profit is more important than the number of units of product A which reflects the wishes of the best customer. He must now express his preferences in a formalized way. He assumes that he can afford to satisfy the wishes of the customer providing

Table 2.3 A subset of Pareto optimal solutions for the production planning problem

Profit decrement (%)	$\bar{x} = [x_1, x_2]^T$	$\bar{f}(\bar{x}) = [f_1(\bar{x}), f_2(\bar{x})]^T$
4	$[\ 88, 112]^T$	$[912, 112]^T$
6	$[101, \ 97]^T$	$[889, \ 97]^T$
8	$[106, \ 90]^T$	$[874, \ 90]^T$
10	$[110, \ 83]^T$	$[855, \ 83]^T$
12	$[116, \ 73]^T$	$[829, \ 73]^T$

that it will cost him less than 8% of the profit he would make if he disregarded these wishes.

The solution which gives the maximal profit is

$$\bar{x}^{0(1)} = [50, 150], \qquad f_1(\bar{x}^{0(1)}) = 950, \qquad f_2(\bar{x}^{0(1)}) = 50$$

The compromise solution which gives 8% of the profit decrement is

$$\bar{x}^* = [106, 90], \qquad f_1(\bar{x}^*) = 874, \qquad f_2(\bar{x}^*) = 90$$

The manager may be unable to determine his preferences by giving the strict values of the function decrement and then he may want to compare solutions for which the values of the profit decrements change in some range, for example between 4% and 12%. In this case the results from Table 2.3 can help him to make the right decision.

Finally consider the beam design problem presented in Fig. 1.3. Assuming that the designer has no idea about the preferences in the criteria, he tries to recognize the situation by comparing the minima and the min–max optimum. The minima are

(1) Volume of the beam
$$\bar{x}^{0(1)} = [159.1, 75.2]^T,$$
$$f(\bar{x}^{0(1)}) = [2.961 \times 10^6, 0.497 \times 10^{-3}]^T$$

(2) Static compliance of the beam
$$\bar{x}^{0(2)} = [0.0, 40.0]^T,$$
$$f(\bar{x}^{0(2)}) = [6.593 \times 10^6, 0.338 \times 10^{-3}]^T$$

The min–max optimum is

$$\bar{x}^* = [237.0, 66.4]^T, \qquad \bar{f}(\bar{x}^*) = [3.7 \times 10^6, 0.425 \times 10^{-3}]^T$$

The designer realizes that he is not satisfied with either solution. To make the right decision he would like to have some representative subset of X^p

Table 2.4   A subset of Pareto optimal solutions for the beam design problem

No.	$\bar{x} = [x_1, x_2]^T$	$f(\bar{x}) = [f_1(\bar{x}), f_2(\bar{x})]^T$
1	$[224.7, 58.6]^T$	$[4.519 \times 10^6, 0.382 \times 10^{-3}]^T$
2	$[235.5, 64.5]^T$	$[3.918 \times 10^6, 0.412 \times 10^{-3}]^T$
3	$[237.1, 68.1]^T$	$[3.522 \times 10^6, 0.438 \times 10^{-3}]^T$
4	$[235.2, 70.2]^T$	$[3.317 \times 10^6, 0.456 \times 10^{-3}]^T$

and let us assume that it would be convenient for him if this subset were such that the values of the objective functions would more or less uniformly cover the set F^p. Such a subset is presented in Table 2.4 and illustrated graphically in Fig. 2.10. Referring to Table 2.4 the designer chooses one solution. This choice is based on his intuition and experience. It may happen, however, that no solution in Table 2.4 seems to be satisfactory and then further exploration of set X^p, presumably in the neighbourhood of a certain solution, would be necessary.

Of course, it is not possible to present all the problems which the decision maker may be faced with. Moreover it is almost impossible to point to any particular method which would solve all these problems. Hence many different methods have been developed. Those which seem to be particularly useful in engineering tasks will be discussed in the next two chapters.

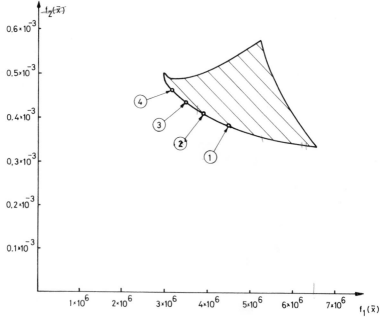

Fig. 2.10   An evenly distributed subset of Pareto optimal solutions for the beam design problem

CHAPTER 3

Methods based on function scalarization

In this chapter we shall describe several basic methods and their variants in which a vector function is transformed to a scalar function. Minimizing this function we may obtain a Pareto optimal solution or a set of such solutions. The forms of transformation differ depending on the method, and influence the results we obtain.

Each method described here requires some information about the preferences in criteria. since the engineer may meet different decision-making problems, and since different information about preferences may be available, it is difficult to indicate which method can be recommended to solve a problem.

The engineer accepts a solution on the basis of the information he has at his disposal. To provide him with as much information as possible, many methods have been recently developed for interactive use. The interactive procedure presents the decision maker, in a series of meetings, with a choice of Pareto optimal solutions which are in some sense representative of all those available. This procedure consists of sequences of decision phases and computation phases. In decision phases the decision-maker decides whether or not a solution is optimal with respect to his implicit preferences. In the latter case he must give some information about the direction in which he expects to obtain a better solution. In the computation phase which follows, the new solution is generated for the next decision phase. The procedure is stopped when the subjectively optimal solution is found. We shall indicate in what way each of the described methods can be used interactively.

3.1 WEIGHTING OBJECTIVES METHOD

The weighting objectives method has received most attention and particular models within this method have been widely applied. The basis of this method consists in adding all the objective functions together using differ-

ent weighting coefficients for each. It means that we change our
multicriterion optimization problem to a scalar optimization problem by
creating one function of the form

$$f(\bar{x}) = \sum_{i=1}^{k} w_i f_i(\bar{x}) \tag{3.1}$$

where $w_i \geqslant 0$ are the weighting coefficients representing the relative importance of the criteria. It is usually assumed that

$$\sum_{i=1}^{k} w_i = 1 \tag{3.2}$$

Since the results of solving an optimization model using (3.1) can vary
significantly as the weighting coefficients change, and since very little is
usually known about how to choose these coefficients, a necessary approach
is to solve the same problem for many different values of w_i. Still, con-
fronted with these solutions, the engineer must then choose among them,
presumably on the basis of his intuition.

Note that the weighting coefficients do not reflect proportionally the
relative importance of the objectives but are only factors which, when
varied, locate points in X^p. For the numerical methods for seeking the
minimum of (3.1) this location depends not only on w_i values, but also on
the units in which the functions are expressed. If we want w_i to reflect
closely the importance of objectives, all functions should be expressed in
units of approximately the same numerical values. We can also transform
(3.1) to the form

$$f(\bar{x}) = \sum_{i=1}^{k} w_i f_i(\bar{x}) c_i \tag{3.3}$$

where c_i are constant multipliers.

The best results are usually obtained if $c_i = 1/f_i^0$. In this case the vector
function is normalized to the following form

$$\bar{f}(\bar{x}) = [\bar{f}_1(\bar{x}), \bar{f}_2(\bar{x}), ..., \bar{f}_k(\bar{x})]^T$$

where $\bar{f}_i(\bar{x}) = f_i(\bar{x})/f_i^0$.

A condition $f_i^0 \neq 0$ is assumed and if it is not satisfied, which rarely happens, the value of c_i must be chosen by the decision maker.

3.1.1 *Example*

Consider the problem illustrated in Fig. 1.4. If (3.1) has the form

$$f(\bar{x}) = w_1 f_1(\bar{x}) + w_2 f_2(\bar{x})$$

Table 3.1 Results of calculations for the normalized weighting method for the beam design problem

No.	w_1	w_2	$x = [x_1, x_2]^T$	$f(x) = [f_1(x), f_2(x)]^T$
1	0.2	0.8	$[207.0, 52.5]$	$[5.101 \times 10^6, 0.362 \times 10^{-3}]^T$
2	0.4	0.6	$[237.1, 68.2]$	$[3.529 \times 10^6, 0.437 \times 10^{-3}]^T$
3	0.6	0.4	$[197.0, 74.1]$	$[2.982 \times 10^6, 0.492 \times 10^{-3}]^T$
4	0.8	0.2	$[163.7, 75.2]$	$[2.948 \times 10^6, 0.499 \times 10^{-3}]^T$

then different values of w_1 and w_2 will always give solutions which are very close to the minimum of the first function. For $x \in X$ the values of $f_1(\bar{x})$ are about ten times greater than the values of $f_2(\bar{x})$ and thus even great improvement in $f_2(\bar{x})$ has very little influence on $f(\bar{x})$. Different values of w_1 and w_2 will give a more representative subset of non-inferior solutions if $f(\bar{x})$ is normalized converting (3.1) to the following form

$$f(\bar{x}) = w_1 \frac{f_1(\bar{x})}{2.961 \cdot 10^6} + w_2 \frac{f_2(\bar{x})}{0.338 \cdot 10^{-3}}$$

Minimizing this function for the different values of w_i we shall obtain the non-inferior solutions presented in Table 3.1 and illustrated graphically in Fig. 3.1.

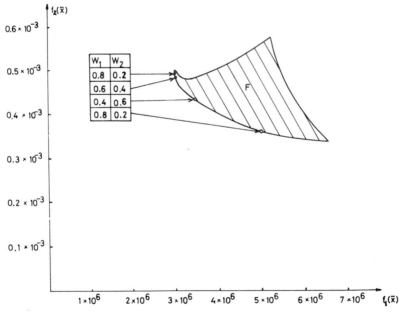

Fig. 3.1 A subset of Pareto optimal solutions obtained for different weighting coefficients

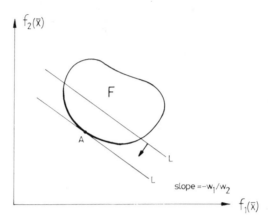

Fig. 3.2 Geometrical interpretation of the weighting objectives method

3.1.2 *Geometrical interpretation*

The weighting objectives method can be interpreted geometrically. Consider the two criterion optimization problem presented in Fig. 3.2. In the space of objectives we can draw a line L with slope $-w_1/w_2$. The set L which represents this line is such that

$$w_1 f_1(\bar{x}) + w_2 f_2(\bar{x}) = c$$

where c is a constant. The minimization of (3.1) can be interpreted as moving the line L with fixed w_1 and w_2 in a positive direction as far as possible from the origin, but keeping the intersection of the sets L and F. The point A for which L is tangent to F will be the minimum of (3.1).

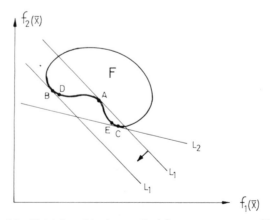

Fig. 3.3 Weighting objectives method for a non-convex problem

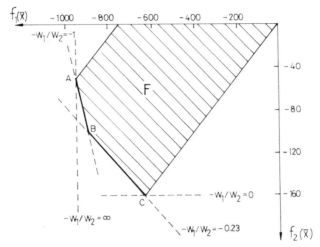

Fig. 3.4 Weighting objectives method for a linear programming problem

Note that for a non-convex problem a great part of the set of non-inferior solutions may not be available, i.e., no values of w_i can locate the points in a certain region of the set F^p. Consider the problem presented in Fig. 3.3. The line L which is tangent at A with slope $-w_1/w_2$ can be moved further in a positive direction until it is tangent at point B. Thus the weighting objectives method with these values of w_i will find point B but not point A. Other values of w_i will find point C. It is easy to see that for this problem the set of non-inferior solutions between points D and E is not available.

For linear programming problems the convexity requirement is satisfied, hence the weighting objectives method may be used to find the complete set of non-inferior solutions. It is impossible, however, to find a representative subset of these solutions. Consider the example presented in Fig. 3.4. For $-w_1/w_2 = -1$ we have an infinite set of solutions (segment AB) which will give the minimum of (3.1). We have the same situation for $-w_1/w_2 = -0.23$. Thus the weighting objectives method cannot locate a solution in the segment AB or BC.

3.1.3 *Similar methods*

More sophisticated methods of this class consider w_i not as coefficients but as functions which reflect the decision maker's preferences. A common one is the additive utility function method for which (3.1.) has the form

$$f(\bar{x}) = \sum_{i=1}^{k} U_i(f_i(\bar{x})) \tag{3.4}$$

where $U_i(f_i(\bar{x}))$ are the utility functions of the criteria.

The utility function method requires that $U_i(f_i(\bar{x}))$ should be known prior to solving the problem. Since the evaluation of $U_i(f_i(\bar{x}))$ for even a simple linear programming problem is difficult, the application of this method is rather limited.

The concept of the utility function is, however, often used in multi-criterion optimization methods. The general form of this function is as follows

$$\max_{\bar{x} \in X_p} U(f_1(\bar{x}), \qquad f_2(\bar{x}), ..., f_k(\bar{x})) \qquad (3.5)$$

where X^p is the set of non-inferior solutions and U is the decision maker's overall utility function. Since the explicit form of (3.5) is not available, in some interactive methods of solution we assume that U exists but is known only implicitly to the decision maker. This means he cannot specify its entire form, but he can answer simple choice questions while comparing two solutions.

The methods which use the utility function concept can ensure the most satisfactory solution to the decision maker provided that U has been correctly assessed and used. The solution is a point at which the set of non-inferior solutions and indifference curves of the decision maker, which can be interpreted as contours of equal utility, are tangent to each other.

The weighting method is designed for use interactively when in each decision phase, values of w_i are generated on the basis of a formal or an informal procedure.

An interactive procedure for linear models has been proposed by Zionts and Wallenius (1976). The first step of the procedure is to choose an arbitrary set of positive multipliers or weights and construct a composite objective function or utility function using these weights. The overall utility function can be unknown explicitly to the decision-maker, but it is assumed to be implicitly a linear function of the objectives. The procedure makes use of such an implicit function on an interactive basis. The utility function is then optimized to obtain a non-inferior solution. For this solution, a subset of efficient variables is selected from the set of non-basic variables. An efficient variable is defined as the one which, when introduced into the basis, cannot increase one objective function without decreasing at least one other objective function.

For each efficient variable a set of trade-offs is defined by which some objective functions are increased and others reduced. This set of trade-offs is presented to the decision maker who is requested to state whether the trade-off is desirable, undesirable or neither. From his answer a new set of weights is constructed and the associated non-inferior solution is found. The process is then repeated by suggesting new trade-offs to the decision maker at each new solution. In case of the linear utility function the procedure guarantees a convergence in a finite number of iterations.

Korhonen and Soismaa (1981) have proposed a modified version of Zionts and Wallenius' method for solving integer linear programming problems.

Note that the concept of the weighting coefficient is also used in other methods. An interactive computer system in which the weighting coefficients are used both for the weighting method and for the min–max method is described in the next chapter.

3.2 HIERARCHICAL OPTIMIZATION METHOD

The hierarchical optimization method has been suggested by Walz (1967) and considers the situation in which the criteria can be ordered in terms of importance. Let the numbering 1 to k reflect this ordering in the sense that the first criterion is the most important and the kth is the least important.

Keeping this order, we minimize each objective function separately, adding in, at each step a new constraint which limits the assumed increase or decrease of the previously considered functions.

This method can be described as follows:

(1) Find the minimum for the primary criterion, i.e., find
$\bar{x}^{(1)} = [x_1^{(1)}x_2^{(1)}, ..., x_n^{(1)}]^T$ such that

$$f_1(\bar{x}^{(1)}) = \min_{\bar{x} \in X} f_1 \bar{x}) \qquad (3.6)$$

Do Step 2 for $i = 2, 3, ..., k$

(2) Find the minimum of the ith objective function, i.e.,
find $\bar{x}^{(i)} = [x_1^{(i)}, ..., x_2^{(i)}, x_n^{(i)}]^T$ such that

$$f_i(\bar{x}^{(i)}) = \min_{\bar{x} \in X} f_i(\bar{x}) \qquad (3.7)$$

with additional constraints

$$f_{j-1}(\bar{x}) \leqslant \left(1 \pm \frac{\epsilon_{j-1}}{100}\right) f_{j-1}(\bar{x}^{(j-1)}) \qquad \text{for } j = 2, 3, ..., i \qquad (3.8)$$

where ϵ_{j-1} are the assumed coefficients of the function increments or decrements given in per cent.

The sign '$+$' refers to the functions which are to be minimized, whereas, '$-$' refers to the functions which are to be maximized. The point $\bar{x}^{(k)} = [x_1^{(k)}, x_2^{(k)}, ..., x_n^{(k)}]^T$ is the optimum determined by this method.

Interactively, the hierarchical optimization method can be run by asking the decision maker to determine, in the ith step, the value of ϵ_i on the basis of the minimum obtained in the ith -1 step. If after all the steps the result is not satisfactory, he may repeat the calculations assuming in each step other values of ϵ_i.

A lexicographic optimization method suggested by Ben-Tal (1980) can be viewed as similar to the hierarchical optimization method. The difference is contained in the assumption that in each step the value of ϵ_i equals zero.

3.2.1 *Example*

The hierarchical optimization method was used in Section 2.4 to explain a decision-making problem for the production planning example (see Fig. 1.3). We assumed that $f_1(\bar{x})$ was more important than $f_2(\bar{x})$. Thus, the first step is to find $\bar{x}^{(1)} = [x^{(1)}, x_2^{(1)}]^T$ such that

$$f_1(\bar{x}^{(1)}) = \min_{x \in X} f_1(\bar{x})$$

In this example we have

$$\bar{x}^{(1)} = [50, 150]^T \quad \text{and} \quad f_1(x^{(1)}) = -950$$

We also assumed that $\epsilon = 8\%$, thus the second step is to find $\bar{x}^{(2)} = [x_1^{(2)}, x_2^{(2)}]^T$ such that

$$f_2(\bar{x}^{(2)}) = \min_{x \in X} f_2(\bar{x})$$

under the additional constraint of the form

$$f_1(\bar{x}) \leqslant \left(1 - \frac{\epsilon_1}{100}\right) f_1(\bar{x}^{(1)})$$

After substitution we have

$$4x_1 + 5x_2 - 874 \geqslant 0$$

The solution is

$$\bar{x}^{(2)} = [106, 90]^T, f_1(\bar{x}^{(2)}) = 874 \quad \text{and} \quad f_2(x^{(2)}) = 106$$

A graphical interpretation of the method for the above calculations is shown in Fig. 3.5.

For this example, the results presented in Table 2.3 illustrate an interactive usage of this method.

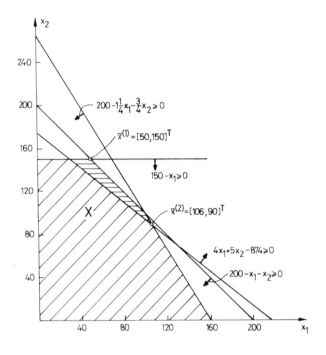

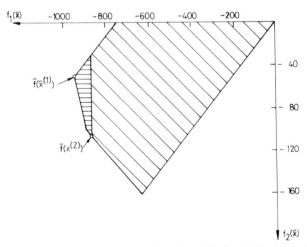

Fig. 3.5 Graphical illustration of the hierarchical optimization method

3.3 TRADE-OFF METHOD

In multicriterion optimization methods the term trade-off is widely used in different contexts. According to *Webster's New World Dictionary of the American Language*, Second College Edition – trade-off means 'an example, especially a giving up of one benefit, advantage, etc. in order to gain another regarded as more desirable'. We shall classify a method in the trade-off category if the concept of trading a value of one objective function for a value of another function is used to determine the next step in seeking the preferred solution.

In most cases the above concept is realized by the minimization of one of the objective functions considering the others as flexible constraints. Thus this method is also called the constraint or ϵ – constraint method.

The simple description of this method is as follows:

(1) Find the minimum of the rth objective function, i.e., find $\bar{x}^*$ such that

$$f_r(\bar{x}^*) = \min_{x \in X} f_r(\bar{x}) \tag{3.9}$$

subject to additional constraints of the form

$$f_i(\bar{x}) \leqslant \epsilon_i \qquad \text{for } i = 1, 2, ..., k \quad \text{and} \quad i \neq r \tag{3.10}$$

where ϵ_i are assumed values of the objective functions which we wish not to exceed.

(2) Repeat (1) for different values of ϵ_i. The information derived from a well chosen set of ϵ_i can be useful in making the decision. The search is stopped when the decision maker finds a satisfactory solution.

It may be necessary to repeat the above procedure for different indices r.

In order to obtain a reasonable choice of ϵ_i it is often useful to minimize each objective function separately, i.e., to find f_i^0 for $i = 1, ..., k$. Knowing these values a more convenient form of (3.10) may be as follows

$$f_i(x) \leqslant f_i^0 + \Delta f_i \qquad \text{for } i = 1, 2, ..., k \text{ and } i \neq r \tag{3,11}$$

where Δf_i are the assumed values of function increments.

It may also be convenient to construct the so-called pay-off table in the following form

	$f_1(\bar{x})$	$f_2(\bar{x}), ..., f_i(\bar{x}), ..., f_k(\bar{x})$		
$\bar{x}^{0(1)}$	f_1^0	$f_{21}, ...,$	$f_{i1}, ...,$	f_{k1}
$\bar{x}^{0(2)}$	f_{12}	$f_2^0, ...,$	$f_{i2}, ...,$	f_{k2}
$\vdots$ $\bar{x}^{0(i)}$	f_{1i}	$f_{2i}, ...,$	$f_i^0, ...,$	f_{ki}
$\bar{x}^{0(k)}$	f_{1k}	$f_{2k}, ...,$	$f_{ik}, ...,$	f_k^0

In this table, row i corresponds to the solution $\bar{x}^{0(i)}$ which minimizes the ith objective function. f_{ji} is the value taken by the jth function $f_j(\bar{x})$ when the ith function $f_i(\bar{x})$ reaches its minimum f_i^0.

The pay-off table may refer to the function increments and then has the form

	$\Delta f_1(\bar{x}) \Delta f_2(\bar{x}), ..., \Delta f_i(\bar{x}), ..., \Delta f_k(\bar{x})$			
$\bar{x}^{0(1)}$	0	$\Delta f_{21}, ...,$	$\Delta f_{i1}, ...,$	Δf_{k1}
$\bar{x}^{0(2)}$	Δf_{12}	$0, ...,$	$\Delta f_{i2}, ...,$	Δf_{k2}
$\vdots$ $\bar{x}^{0(i)}$	Δf_{1i}	$\Delta f_{2i}, ...,$	$0, ...,$	Δf_{ki}
$\bar{x}^{0(k)}$	Δf_{1k}	$\Delta f_{2k}, ...,$	$\Delta f_{ik}, ...,$	0

In this table a Δf_{ji} is defined as follows

$$\Delta f_{ji} = |f_j(\bar{x}^{0(i)}) - f_j^0| \qquad (3.12)$$

Similarly we may construct the pay-off table which refers to relative increments of the functions.

The pay-off table in all these forms may be useful in making the decision while applying other multicriterion optimization methods.

3.3.1 *Example*

Consider the metal cutting problem presented in Fig. 1.6. Let us choose $f_1(\bar{x})$ as the function to be minimized and formulate the corresponding scalar optimization problem.

Find $\bar{x}^* = [x_1, x_2]^T$ such that

$$f_1(\bar{x}^*) = \min_{x \in X} - 6000 x_1 x_2$$

subject to the additional constraint

$$- \frac{1.28 \times 10^7}{x_1^{3.33} x_2^{2.22}} \leqslant - \epsilon_2$$

If it is difficult to assume *a priori* the values of ϵ_2 then the pay-off table will provide the information as to the possible range of ϵ_2. In our example the pay-off table for the function values is

	$f_1(\bar{x})$	$f_2(\bar{x})$
$\bar{x}^{0(1)} = [94.9, 1.0]^T$	$-5.697 \cdot 10^5$	-3.326
$\bar{x}^{0(2)} = [179.9, 0.05]^T$	$-0.538 \cdot 10^5$	-307.4

For the function increments we have

	$\Delta f_1(\bar{x})$	$f_2(\bar{x})$
$\bar{x}^{0(1)}$	0	-304.0
$\bar{x}^{0(2)}$	$-5.159 \cdot 10^5$	0

and for the relative function increments we have

	$z_1(\bar{x})$	$z_2(\bar{x})$
$\bar{x}^{0(1)}$	0	91.412
$\bar{x}^{0(2)}$	9.576	0

Table 3.2 Results of calculations for the trade-off method for the metal cutting problem

No.	ϵ_2	$\bar{x} = [x_1, x_2]^T$	$f_1(\bar{x})$
1	45	$[43.4, 1.0\ ]^T$	2.60×10^5
2	60	$[40.0, 1.0\ ]^T$	2.40×10^5
3	75	$[48.9, 0.67]^T$	1.96×10^5
4	90	$[57.6, 0.48]^T$	1.65×10^5

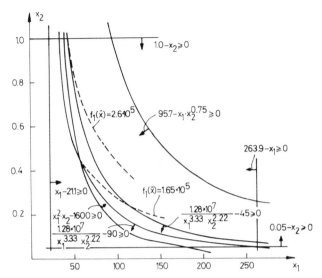

Fig. 3.6 Graphical illustration of the trade-off method

Now the engineer must decide for which values of $\epsilon_2, f_1(x)$ is minimized. Assuming that these values are 45, 60, 75, 90 we obtain the results as presented in Table 3.2. Graphical solution of this problem for $\epsilon_2 = 45$ and $\epsilon_2 = 90$ is presented in Fig. 3.6.

The trade-off method in the form described above usually requires a long computation time and for more than three criteria becomes unwieldy.

In spite of this engineers are inclined to use this method, especially when they hesitate while building a model, i.e., when they are not sure whether to classify certain quantities as objective functions and others as constraints.

3.3.2 Variants of the trade-off method

There are several versions of the trade-off method whose aim is to make the search for a preferred solution more efficient. We shall describe some of them briefly. Haimes and Hall (1974) have suggested the surrogate worth trade-off method in which objective trade-offs are used as the information carrier and the decision maker responds by expressing his degree of preference over the prescribed trade-offs by assigning numerical values to each surrogate worth function. An interactive procedure is such that the values of either the surrogate worth function of the marginal rates of substitution are used to determine the direction in which the decision maker must assess his preference at each trial solution in order to determine the step size.

Such a requirement can be inconvenient for the decision maker since he usually does not know the explicit form of his utility function. To avoid this disadvantage Oppenheimer (1978) has proposed the method in which the local proxy preference function is updated at each iteration by assessing a new marginal rate of substitution vector. Then the proxy function is maximized to find a better point. This method, however, does not guarantee that a Pareto optimal solution will be obtained in each iteration and the systematic procedure to maximize the proxy functions has not been established for this method. In order to improve this method, Sakawa (1981) has proposed the so-called sequential proxy optimization technique. In this interactive on-line procedure the values of the marginal rates of substitution assessed by the decision maker are used to determine the direction to which the utility function increases most rapidly and the local proxy preference function is updated to determine the optimal step size. For the concave and strictly decreasing utility function and for the convex optimization problem the procedure guarantees that a Pareto optimal solution will be obtained.

A slightly different trade-off cut approach has been proposed by Musselman and Tavalage (1980). This approach concentrates on reducing the feasible region in decision space through the use of trade-off cuts.

In supplying trade-offs between the objectives, the decision maker establishes a cutting plane in the objective space which removes from further consideration large areas of the decision space. Although the decision maker is not aware of the extent to which he has reduced his decision space, he is assured that the preferred solution has not been carelessly eliminated. The cutting plane established from these trade-offs focuses on the most relevant portion of the decision space, but the accomplishment of this is based more on eliminating the inferior portion of the space than on identifying the superior portion. Basically, the cut acts as a restriction on where not to proceed.

Two disadvantages of the methods described here are that

(i)   they cannot be used to solve non-convex problems, and
(ii) they allow a satisfactory solution to be found only in a certain region of Pareto optimal solutions, but do not provide a general outlook on the possible range of objectives and thus the final decision is influenced by the starting point chosen,

3.4 GLOBAL CRITERION METHOD

In this method an optimal solution is a vector of decision variables which minimizes some global criterion. A function which describes this global criterion is a measure of 'how close the decision maker can get to the ideal vector $\bar{f}^{0}$. The most commmon form of this function is

$$f(\bar{x}) = \sum_{i=1}^{k} \left(\frac{f_i^0 - f_i(\bar{x})}{f_i^0} \right)^P \qquad (3.13)$$

For this formula Boychuk and Ovchinnikov (1973) have suggested $p = 1$, and Salukvadze (1974) has suggested $p = 2$, but other values of p can also be used. Naturally, the solution obtained after minimizing (3.13) will differ greatly according to the value of p chosen. Thus the problem is to determine which p would result in a solution that is the most satisfactory for the decision maker. It is also possible that whatever p is chosen formula (3.13) might give a solution that is unacceptable to the decision maker.

3.4.1 *Example*

Consider the production planning problem presented in Fig. 1.3. For this problem the ideal solution gives $\bar{f}^0 = [-950, -160]^T$. The global function for $p = 1$ is

$$f(\bar{x})_{p=1} = 2 - 0.01064x_1 - 0.00526x_2$$

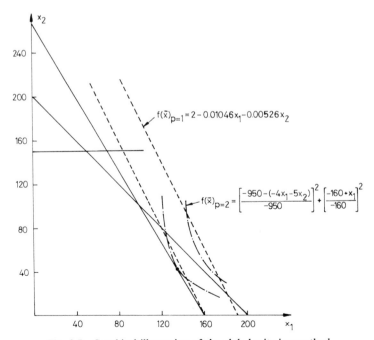

Fig. 3.7 Graphical illustration of the global criterion method

The minimum of this function gives the solution $\bar{x} = [160, 0]^T$, for which $\bar{f}(\bar{x}) = [-640, -160]^T$.

The global function for $p = 2$ is

$$f(\bar{x})_{p=2} = \left[\frac{-950 - (-4x_1 - 5x_{2)}}{-950} \right]^2 + \left[\frac{-160 - (-x_1)}{-160} \right]^2$$

The minimum of this function gives the solution

$$\bar{x} = [135.1, 41.4]^T \qquad \text{for which } \bar{f}(\bar{x}) = [-747.4, -135.1]^T$$

The graphical illustration of this example for $p = 1$ and $p = 2$ in the space of decision variables is presented in Fig. 3.7.

Note that if $p \geqslant 2$ linear models cannot be solved by the simplex method of linear programming since (3.13) becomes a non-linear function.

3.4.2 Other forms of the global function

Another possible measure of 'closeness to the ideal solution' is a family of L_p-metrics defined as follows

$$L_p(f) = \left[\sum_{i=1}^{k} |f_i^0 - f_i(x)|^p \right]^{1/p}, \qquad 1 \leqslant p \leqslant \infty \qquad (3.14)$$

For example

$$L_1(f) = \sum_{i=1}^{k} |f_i^0 - f_i(\bar{x})|$$

$$L_2(f) = \left[\sum_{i=1}^{k} (f_i^0 - f_i(\bar{x}))^2 \right]^{1/2} \qquad (3.15)$$

$$L_\infty(f) = \max_{i \in I} |f_i^0 - f_i(\bar{x})|$$

Note that the minimization of $L_2(f)$ leads to the minimization of the Euclidean distance between $\bar{x} \in X$ and the ideal solution.

Instead of deviation in an absolute sense it is recommended to use relative deviations in (3.14) such as

$$\frac{f_i^0 - f_i(x)}{f_i^0}$$

which have a substantive meaning in any given context. The relevant

L_p-metrics are

$$L_p(f) = \left[\sum_{i=1}^{k} \left| \frac{f_i^0 - f_i(\bar{x})}{f_i^0} \right|^p \right]^{1/p}, \qquad 1 \leqslant p \leqslant \infty \qquad (3.16)$$

Koski (1981) has suggested L_p-metrics with a normalized vector objective function of the form

$$\bar{\bar{f}}_i(x) = \frac{f_i(\bar{x}) - \min_{x \in X} f_i(\bar{x})}{\max_{x \in X} f_i(\bar{x}) - \min_{x \in X} f_i(\bar{x})}$$

In this case the values of every normalized function are limited to the range $[0, 1]$.

The global criterion method with $L_\infty(f)$ metric is also called the min–max method since for this metric the optimum $\bar{x}^*$ is defined as follows

$$f(\bar{x}^*) = \min_{x \in X} \max_{i \in I} \left| \frac{f_i^0 - f_i(\bar{x})}{f_i^0} \right| \qquad (3.17)$$

The above formula is the same as the first step of (2.18) which usually defines the optimum in the min–max sense.

Using the global criterion method one non-inferior solution is obtained. If certain parameters w_i are used as weights for the criteria a required set of non-inferior solutions can be found. This and other possibilities resulting from applying methods based on the min–max approach will be discussed in Chapter 4.

A slightly different method has been suggested by Wierzbicki (1978), (1980). In this method the global function has a form such that it penalizes the deviations from the so-called reference objective. Any reasonable or desirable point in the space of objective chosen by the decision maker can be considered as the reference objective.

Let $\bar{f}^r = [f_1^r, f_2^r, ..., f_k^r]^T$ be a vector which defines this point. Then the function which is minimized has the form

$$P(\bar{x}, \bar{f}^r) = -\sum_{i=1}^{k} (f_i(\bar{x}) - f_i^r)^2 + \varrho \sum_{i=1}^{k} (\max(0, f_i(\bar{x}) - f_i^r)^2) \qquad (3.18)$$

where $\varrho > 0$ is a penalty coefficient which in this method can be chosen as constant.

Minimizing (3.18) for the assumed point $\bar{f}^r$ we obtain the non-inferior solution which is close to this point. If for different points $\bar{f}^r$ the procedure is carried out, some representation of non-inferior solutions can be found.

Note that if numerical methods of minimization are used, the units in which the objective functions are expressed will also have an influence on the results obtained.

3.5 GOAL PROGRAMMING METHOD

Goal programming was proposed by Charnes and Cooper (1961) and Ijiri (1965) for a linear model. This method requires the decision maker to specify goals for each objective that he wishes to attain.

The goals, i.e., their quantitative values, are considered as additional constraints in which new variables are added to represent deviations from the predetermined goals. The objective function specifies the deviations from these goals and priorities for the achievement of each goal, in quantitative terms.

The most common form of goal programming formulation is as follows. Find $\bar{x}^* = [x_1^*, x_2^*, ..., x_n^*]^T$ such that

$$\min a = \{P_1 h_1(d^-, d^+), P_2 h_2(d^-, d^+), ..., P_1 h_1(d^-, d^+)\} \qquad (3.19)$$

subject to

$$g_j(\bar{x}) + d_j^- - d_j^+ = b_j \qquad j = 1, 2, ..., m \qquad (3.20)$$

$$f_i(\bar{x}) + d_i^- - d_i^+ = b_i \qquad i = 1, 2, ..., k \qquad (3.21)$$

$$\bigwedge_i (d_i^-, d_i^+ \geq 0)$$

$$\bigwedge_i (d_i^-, d_i^+ = 0)$$

where

b_i = quantitative value of the ith goal,

d_i^- = negative deviation from the ith goal,

d_i^+ = positive deviation from the ith goal,

$h_i(d^-, d^+)$ = function of the deviational variables called the ith achievement functions, where $i = 1, 2, ..., l$,

P_i = priority coefficient for the ith achievement function.

Equations (3.20) and (3.21) represent the desired goals b_i and the functions which effect these goals. Goal deviation variables are added to this set of equations to force equality between the desired goals and the functions which make up the goals.

The specification of the achievement function (3.19) is the key element for the practical deployment of this method. In this function it is assumed that $P_i \gg P_{i+1}$, which means that no number N, however large, can make $NP_{i+1} > P_i$. This property of goal programming allows the absolute ordering of goals.

The solution algorithm for goal programming is as follows.

(1) Find the solution that minimizes the first achievement function, i.e., the function with priority level 1

$$h_1^*(d^-, d^+) = \min_{\bar{x} \in X} h_1(d^-, d^+) \qquad (3.22)$$

(2) Do Step 3 for $j = 2, 3, ..., l$

(3) Find the solution that minimizes the jth achievement function, i.e.,

$$h_j^*(d^-, d^+) = \min_{\bar{x} \in X} h_j(d^-, d^+) \qquad (3.23)$$

subject to additional constraints of the form

$$h_{i-1}(d^-, d^+) \leqslant h_{i-1}^*(d^-, d^+) \qquad \text{for } i = 1, 2, ..., j \quad (3.24)$$

In other words we minimize the second and the following achievement functions but under no circumstances can the previously considered achievement functions be greater than their minima.

(4) The solution determined while minimizing $h_l(d^-, d^+)$ is the optimum.

3.5.1 *Example*

Consider the production planning problem illustrated in Fig. 1.3. Let us assume that the priorities are:

P_1 The constraints given by (1.15) must be satisfied.
P_2 Produce at least 100 units of product A which means that we want to satisfy the best customer's order by producing for him at least 100 units.
P_3 Try to obtain the profit as close as possible to $1000

Now the goal programming model is

Minimize

$$(d_1^+ + d_2^+ + d_3^+), (d_5^-), (d_4^-)$$

subject to

$$G_1 : x_1 + x_2 + d_1^- - d_1^+ = 200$$

$$G_2 : 1.25x_1 + 0.75x_2 + d_2^- - d_2^+ = 200$$

$$G_3 : x_2 + d_3^- - d_3^+ = 150$$

$$G_4 : 4x_1 + 5x_2 + d_4^- - d_4^+ = 1000$$

$$G_5 : x_1 + d_5^- + d_5^+ = 100$$

$$x, d^-, d^+ \geqslant 0$$

The graphical illustration of the above model is shown in Fig. 3.8. The achievement function consists of the priorities mentioned above. To satisfy the first priority we minimize $(d_1^+ + d_2^+ + d_3^+)$ as G_1, G_2 and G_3 are physical constraints. The priority is satisfied when $d_1^+ = d_2^+ = d_3^+ = 0$. Otherwise the point x will be outside the set of feasible solutions which in Fig. 3.8 is denoted by the shaded area. The second priority is to produce 100 units of product A which means the minimization of d_5^-. We can fully satisfy this goal because we can have $d_5^- = 0$. For the third priority the set of the feasible solutions is limited to the cross shaded area. Finally to achieve the profit

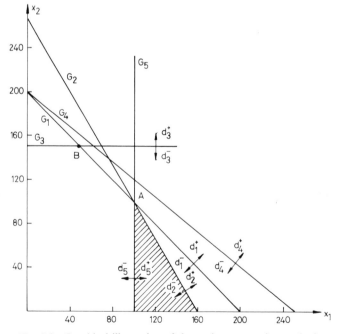

Fig. 3.8   Graphical illustration of the goal programming method

maximization goal of the priority level 3 we must minimize d_4^-. We may proceed until we reach point A without degrading the solution to priority level 2. Hence the point A is the final solution for which

$$\bar{x} = [100, 100]^T \qquad a = \{0, 0, 100\} \qquad \bar{f} = [900, 100]^T$$

Note that if the achievement function has the form

$$\min a = \{(d_1^+ + d_2^+ + d_3^+), (d_4^-), (d_5^-)\}$$

then the final solution will be point B for which

$$\bar{x} = [50, 150]^T \qquad a = \{0, 50, 50\} \qquad \bar{f} = [950, 50)$$

Goal programming has been developed mainly for linear models and some special computer programs for these models are available (see Lee (1972) and Zeleny (1974)).

Interactively, the goal programming method is used by a weight w_i being prescribed to each achievement function. For varying values of w_i for $i = 1, 2, ..., l$ different solutions can be obtained. This interactive algorithm has been proposed by Dyer (1972) in which at each step the decision maker must provide information regarding his local trade-offs between criteria at specific points.

CHAPTER 4

Methods based on min–max approach

In the previous chapter we have discussed the methods in which some form of a scalar function is used to solve a vector optimization problem. This chapter will be devoted to the methods which are developed on the basis of the min–max approach. These methods provide the engineer with both the best compromise solution and a desirable set of Pareto optimal solutions.

First we shall describe two algorithms for comparing solutions which select a set of Pareto optimal solutions and the min–max optimum. These algorithms can be implemented for any multicriterion optimization method which generates a set of feasible solutions. Two random search methods are presented here. Later we shall see that usually the min-max approach can be considered as a scalar optimization problem and then most of the well known sequential methods can be used to find the optimum.

Next we shall discuss how the min–max approach can be used to obtain an evenly distributed set of Pareto optimal solutions, a required set of solutions which are in the neighbourhood of a desirable solution, and a solution which fulfils the assumed goals for which priorities may be assumed. Used interactively this approach provides the engineer with a number of solutions for any conflicting situation and thus enable him to find a satisfactory solution in an efficient way.

Finally we shall describe an interactive computer system which incorporates several multicriterion methods.

4.1 TWO ALGORITHMS FOR COMPARING SOLUTIONS

Let us assume that we have a method which generates a set of feasible solutions. For multicriterion optimization we would like to select from this set, a subset which contains only Pareto optimal solutions, and the min–max optimal solution. We shall describe two algorithms called Pareto and Min–max respectively which enable us to make this selection.

In order to solve a problem the engineer is often forced to prepare a problem-oriented method, usually a heuristic one, which gives only a finite set of feasible solutions to be compared. Both algorithms can be used with the above method. In the next section we shall apply these algorithms to a simple random search method.

4.1.1 *Pareto algorithm*

The algorithm which selects the Pareto optimal solutions from a given set of feasible solutions has been proposed by Gerlach (1980) and is based on the contact theorem which is one of the main theorems in multicriterion optimization.

First let us define a negative cone. The negative cone in E^k is the set

$$C^- = \{\bar{f} \in E^k \,|\, \bar{f} \leqslant 0\} \qquad (4.1)$$

Thus the contact theorem is:

A vector f^* is a Pareto optimal solution for the multicriterion optimization problem if and only if

$$(C^- + \bar{f}^*) \cap F = \{\bar{f}^*\} \qquad (4.2)$$

For the proof see Lin (1976).

A graphical illustration of this theorem for a two criterion problem is presented in Fig. 4.1.

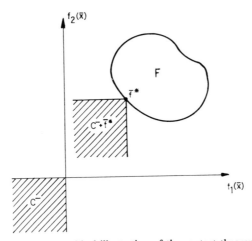

Fig. 4.1 Graphical illustration of the contact theorem

Consider two solutions $\bar{x}^{(1)}$ and $\bar{x}^{(2)}$ for which we may have two specific cases

(1) $(C^- + \bar{f}(\bar{x}^{(1)})) \subset (C^- + \bar{f}(\bar{x}^{(2)}))$ (4.3)

(2) $(C^- + \bar{f}(\bar{x}^{(1)})) \supset (C^- + f(x^{(2)}))$ (4.4)

A graphical illustration of these cases is presented in Fig. 4.2. We denote

$\bar{x}^{(l)} = [x_1^{(l)}, x_2^{(l)}, ..., x_n^{(l)}]^T = $ any given point in X,

$f(\bar{x}^{(l)}) = [f_1(\bar{x}^{(l)}), f_2(\bar{x}^{(l)})), ..., f_k(\bar{x}^{(l)})]^T = $ vector of objective functions for the point $\bar{x}^{(l)}$,

$\bar{x}_j^p = [x_{1j}^p, x_{2j}^p, ..., x_{nj}^p]^T = $ the jth Pareto optimal solution,

$\bar{f}_j^p = [f_{1j}^p, f_{2j}^p, ..., f_{kj}^p]^T = $ vector of objective functions for the jth Pareto optimal solution.

Now the problem is to choose from any given set of solutions

$L = \{1, 2, ..., l, ..., l^a\}$, the set of Pareto optimal solutions
$J = \{1, 2, ..., j, ..., j^a\}$,

The main concept of the Pareto algorithm is as follows. Let $\bar{x}^{(l)}$ be a new solution to be considered. If in the set of Pareto optimal solutions there is a solution x_j^p such that it

(i) satisfies (4.3) then $\bar{x}^{(l)}$ is substituted for $\bar{x}_j^p$, or
(ii) satisfies (4.4) then $\bar{x}^{(l)}$ is discarded.

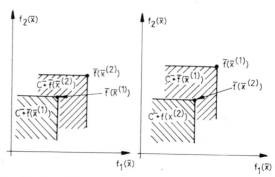

Fig. 4.2 Graphical illustration of (4.3) and (4.4) formulae

If none of the solutions from the set of Pareto optimal solutions satisfies either (4.3) or (4.4), then $\bar{x}^{(l)}$ becomes a new Pareto optimal solution.

The steps of the algorithm can be described in detail as follows:

(1) Read k, n, l^a.

(2) Set $f_{i1}^p = \infty$ for $i = 1, 2, ..., k$ and $j^a = 1$.

(3) Set $l = 1$.

(4) Read $\bar{x}^{(l)}$ and $f(\bar{x}^{(l)})$.

(5) Set $j = 1$.

(6) If for every $i \in I$ we have $f_i(\bar{x}^{(l)}) < f_{ij}^p$ then substitute $\bar{x}_j^p = \bar{x}^{(l)}$ and $f_j^p = f(\bar{x}^{(l)})$ and go to 10, otherwise go to 7.

(7) If for every $i \in I$ we have $f_i(\bar{x}^{(l)}) > f_{ij}^p$ then go to 10 otherwise go to 8.

(8) Set $j = j + 1$.

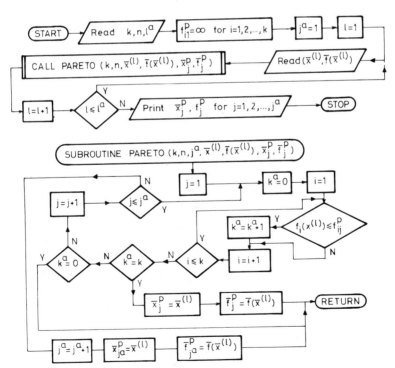

Fig. 4.3 A detailed flow diagram for the PARETO algorithm

(9) If $j > j^a$ then $j^a = j^a + 1$ and $x^p_{ja} = \bar{x}^{(l)}$ and $\bar{f}^p_j = \bar{f}(\bar{x}^{(l)})$ and go to 10, otherwise go to 6.

(10) Set $l = l + 1$.

(11) If $l \leqslant l^a$, then go to 4, otherwise go to 12.

(12) Print $\bar{x}^p_j$ and $\bar{f}^p_j$ for $j = 1, 2, ..., j^a$.

A detailed flow diagram of the algorithm described above is presented in Fig. 4.3. The essential part of this algorithm is contained in subroutine PARETO with dummy arguments $k, n, \bar{x}^{(l)}, \bar{f}(\bar{x}^{(l)}), \bar{x}^p_j, \bar{f}^p_j$, whose actual values are read and printed in the main program. Subroutine PARETO can be used with any method which generates $\bar{x}^{(l)}$ and $\bar{f}(\bar{x}^{(l)})$ for the problem originally stated in (2.8). In Appendix A a program in FORTRAN is shown for this subroutine.

4.1.2 *Min–max algorithm*

The aim of this algorithm is to choose from any given set of solutions $L = \{1, 2, ..., l, ..., l^a\}$, the min–max optimal solution. We assume that the ideal vector $\bar{f}^0$ is given.

The steps of the algorithm are as follows:

(1) Read $k, n, l^a, \bar{f}^0$.

(2) Set $v^*_1 = \infty$.

(3) Set $l = 1$.

(4) Read $\bar{x}^{(l)}$ and $\bar{f}(\bar{x}^{(l)})$.

(5) Evaluate vector $\bar{z}(\bar{x}^{(l)})$ using formula (2.17).

(6) If $\bar{z}(\bar{x}^{(l)}) = 0$ then retain this solution as the optimum since there is no better solution, and go to 11, otherwise go to 7.

(7) Find the maximal values of all the sreps of formula (2.18) for the point $\bar{x}^{(l)}$. These values are denoted v_r for $r = 1, 2, \ldots, k$, and can be evaluated as follows

$$v_1 = \max_{i \in I} \{z_i(x^{(l)})\}$$

and then $I_1 = \{i_1\}$, where i_1 is the index for which the value of $z_i(\bar{x}^{(l)})$ is maximal,

$$v_2 = \max_{\substack{i \in I \\ i \notin I_1}} \{z_i(\bar{x}^{(l)})\}$$

and the $I_2 = \{i_1, i_2\}$, where i_2 is the index for which the value of $z_i(\bar{x}^{(l)})$ is maximal,

$$v_r = \max_{\substack{i \in I \\ i \notin I_{r-1}}} \{z_i(\bar{x}^{(l)})\}$$

and then $I_r = \{I_{r-1}, i_r\}$, where i_r is the index for which the value of $z_i(\bar{x}^{(l)})$ is maximal,

$$v_k = z_i(\bar{x}^{(l)}) \qquad \text{for } i \in I \text{ and } i \notin I_{k-1}$$

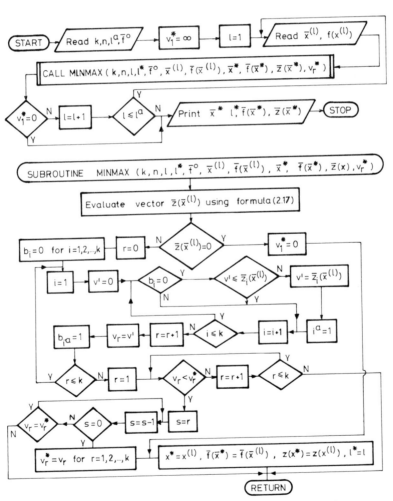

Fig. 4.4 A detailed flow diagram for the MINMAX algorithm

(8) Replace v_r^* by v_r for $r = 1, 2, \ldots, k$ and retain this solution as the optimum if the following sentential function is satisfied

$$v_1 < v_1^* \underset{r \in \{2, \ldots, k\}}{\vee} ((v_r < v_r^*) \underset{s \in \{1, \ldots, r\}}{\wedge} (v_s = v_s^*)) \qquad (4.5)$$

where $v_1^*, v_2^*, \ldots, v_k^*$ is the set of optimal values of relative increments ordered non-increasingly.

(9) Set $l = l + 1$.

(10) If $l \leqslant l^a$ then go to 4, otherwise go to 11.

(11) Print $x^*, l^*, \bar{f}(\bar{x}^*), \bar{z}(\bar{x}^*)$.

A detailed flow diagram of the algorithm described above is presented in Fig. 4.4. As we can see the structure of this flow diagram is the same as for the Pareto algorithm. The subroutine MINMAX is considered separately in order to use it with any method which generates $\bar{x}^{(l)}$ and $\bar{f}(\bar{x}^{(l)})$. For this subroutine a program in FORTRAN is also given in Appendix A.

4.2 METHODS FOR SEEKING THE MIN-MAX OPTIMUM

Two basic classes of function minimization methods can be distinguished: (i) exploratory methods and (ii) sequential methods.

4.2.1 Exploratory methods

In exploratory methods a point is generated by means of a rule which disregards the results previously obtained. The simplest rule is contained in the systematic search method in which we set up a grid with points spaced together closely enough to define a minimum and we calculate the value of the objective function for each grid point. The minimum of the function is determined by inspection. Exploratory methods can be used to solve problems with only few decision variables and are less precise than sequential methods, but all the classes of non-linear programming problems can be solved using these methods.

A good example of exploratory methods in the Monte Carlo method in which a certain number of points are picked at random over the estimated range of all the variables. This may be done formally by obtaining the randomly selected value for x_i from the following formula

$$x_i = x_i^a + \delta_i(x_i^b - x_i^a) \qquad \text{for } i = 1, 2, \ldots, n \qquad (4.6)$$

where x_i^a is the estimated or given lower limit for x_i, x_i^b is the estimated or given upper limit for x_i, and δ_i is a random number between zero and one.

If we decide to generate the values of variables for l^a points, we generate random numbers δ_i for each point and use (4.6) to obtain the values of the variables x_i. We test each generated point for violation and discard it if it is not a feasible solution. If the point is in the feasible region, we evaluate the objective function for that point. The best result is taken as the minimum.

A new set of random numbers is generated for each of l^a points. Now we shall describe two Monte Carlo methods used for finding the min-max optimum.

Method 1
This method is based upon double exploration of the area of the space of variables, first for seeking the ideal vector $\bar{f}^0$, second for seeking the min-max optimum. This can be described as follows:

Do steps 1, 2, 3, 4, for $l = 1, 2, \ldots, l^a$

(1) Generate a random point $\bar{x}^{(l)}$.

(2) If the point $\bar{x}^{(l)}$ is not in the feasible region go to 1, otherwise go to 3.

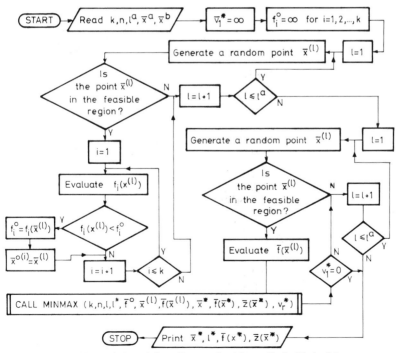

Fig. 4.5 A general flow diagram for Monte Carlo Method 1

(3) Evaluate $f_i(\bar{x}^{(l)})$ for $i = 1, 2, \ldots, k$

(4) Replace f_i^0 by $f_i(\bar{x}^{(l)})$ for every i for which $f_i(\bar{x}^{(l)}) < f_i^0$

Do steps 5, 6, 7, 8, for $l = 1, 2, \ldots, l^a$

(5) Generate a random point $\bar{x}^{(l)}$.

(6) If the point $x^{(l)}$ is not in the feasible region go to 5, otherwise go to 7.

(7) Evaluate $f_i(\bar{x}^{l})$ for $i = 1, 2, \ldots, k$.

(8) Call subroutine MINMAX for checking if the point $\bar{x}^{(l)}$ is the min-max optimum.

The general flow diagram for this method is presented in Fig. 4.5.

Method 2
In this method we explore the area of the space of variables only once, and during this exploration we create a set of Pareto optimal solutions and seek for the ideal vector f^0. Next we look through this set and check which solution is the min-max optimum. This can be described as follows:

Do steps 1, 2, 3, 4, 5, for $l = 1, 2, \ldots, l^a$

(1) Generate a random point $\bar{x}^{(l)}$.

(2) If the point $\bar{x}^{(l)}$ is not in the feasible region go to 1, otherwise to 3.

(3) Evaluate $f_i(\bar{x}^{(l)})$ for $i = 1, 2, \ldots, k$.

(4) Replace f_i^0 by $f_i(\bar{x}^{(l)})$ for every i for which $f_i(x^{(l)}) < f_i^0$.

(5) Call subroutine PARETO for checking if the point $\bar{x}^{(l)}$ is Pareto optimal.

Do steps 6, 7 for $j = 1, 2, \ldots, j^a$.

(6) Evaluate $f_i(\bar{x}_j^p)$ for $i = 1, 2, \ldots, k$.

(7) Call subroutine MINMAX for checking if the point x_j^p is the min-max optimum.

The general flow diagram for this method is presented in Fig. 4.6.
Method 2, compared with Method 1, may use less computer time (we explore the space of variables only once) but it takes much more computer memory as the whole set of Pareto optimal solutions has to be stored. Of

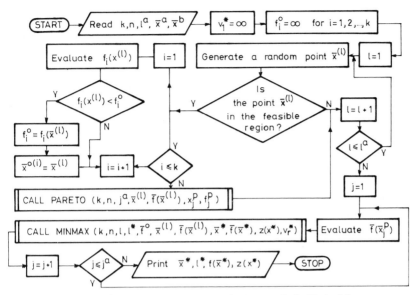

Fig. 4.6   A general flow diagram for Monte Carlo Method 2

course, we may want to know this set in order to make the right decision, but if the set is large it may be difficult to look through all the solutions. We may reduce this set by introducing the additional constraints of the form

$$f_i(\bar{x}) \leqslant f_i^0 \qquad \text{for } i = 1, 2, \ldots, k \qquad (4.7)$$

where values of f_i^0 are chosen by the decision maker. This approach is similar to the ideal vector displacement method described in the next section.

Method 2 may be more effective for problems with a large number of constraints because then we may expect that the number of Pareto optimal solutions will be relatively small. This number will usually be small also for discrete programming problems.

If a computer graphic output is available, Method 2 provides also the possibility of having a graphical illustration of the Pareto optimal solutions for two criterion problems.

Note that while lacking the precision of the sequential methods, both methods described above impose no restriction on a problem to be solved.

In Appendix A a FORTRAN program is given for both methods.

4.2.2 *Sequential methods*

In sequential methods a point is established on the basis of the previously obtained results, which indicate where the minimum is likely to be or the general direction in which it is likely to be. The procedure is then repeated and each time the new point thus established is closer to the minimum. These methods are generally much more efficient and thus much more highly developed, but they are designed to solve only continuous convex problems.

As we saw in Section 2.3 the min-max optimum is defined by a recurrence formula. Thus it is impossible to use it in its full form with single criterion sequential methods. However, in most problems it is enough to refer to the first step of formula (2.18). Then a function which is to be minimized has the form

$$v(\bar{x}) = \max_{i \in I} \{ z_i(\bar{x}) \} \tag{4.8}$$

and the optimization problem is to find $x^* \in X$ such that

$$v(\bar{x}^*) = \min_{x \in X} \max_{i \in I} \{ z_i(\bar{x}) \} \tag{4.9}$$

Note that for convex problems the above formula always defines the min-max optimum.

For non-linear models many single criterion optimization methods and computer programs accept the above form of the minimized function. The author has adapted and successfully used the following methods:

(1) Hooke and Jeeves (1961) direct search method,

(2) Fletcher and Powell's (1963) variable metric method,

(3) the flexible tolerance method (Himmelblau (1972)).

For the first two methods a penalty function approach handles models with constraints.

Note that the problem of seeking the minimum of (4.8) can be sequentially replaced by

$$v(\bar{x}^*) = \min_{\bar{x} \in X} x_0 \quad \text{and} \quad x_0 = \max_{i \in I} \{ z_i(\bar{x}) \} \tag{4.10}$$

$$v(\bar{x}^*) = \min_{\bar{x} \in X} x_0 \quad \text{and} \quad x_0 \geqslant \max_{i \in I} \{ z_i(\bar{x}) \} \tag{4.11}$$

$$v(\bar{x}^*) = \min_{\bar{x} \in X} x_0 \quad \text{and} \quad x_0 \geqslant z_i(\bar{x}) \qquad \text{for } i = 1, 2, \ldots, k \qquad (4.12)$$

where x_0 is an additional variable.

Hence the problem of seeking the min-max optimum can be transformed to the problem of seeking the minimum of x_0 under the following additional constraints

$$x_0 \geqslant z_i(\bar{x}) \qquad \text{for } i = 1, 2, \ldots, k \qquad (4.13)$$

This becomes important for linear models because the constraints (4.13) may be linear functions and thus the simplex algorithm can be used to find the min-max optimum. To satisfy the linearity condition we have to evaluate the values of $z_i(\bar{x})$ for the minimized functions from

$$z_i(\bar{x}) = \frac{f_i(\bar{x}) - f_i^0}{f_i^0} \qquad (4.14)$$

and for the maximized functions from

$$z_i(\bar{x}) = \frac{f_i^0 - f_i(\bar{x})}{f_i^0} \qquad (4.15)$$

Consider the example presented in Fig. 1.3. The problem of seeking the min-max optimum can be written as follows:

Find $\bar{x}^* = [x_0^*, x_1^*, x_2^*]^T$ such that

$$v(\bar{x}^*) = \min_{\bar{x} \in X} x_0$$

under two additional constraints

$$x_0 \geqslant \frac{2}{475} x_1 + \frac{1}{190} x_2 - 1$$

$$x_0 \geqslant \frac{x_1}{160} - 1$$

The above constraints are linear functions of variables x_0, x_1, x_2.

It is clear that for linear models the function (4.8) can also be used with well developed single criterion methods for models with linear constraints and non-linear objective function (see, for instance, Zoutendijk (1974) and Goncalves (1974)).

Exploratory and sequential methods can be used together. For example in the interactive computer system described in Section 4.4, for which a computer program is provided in Appendix B, one of the methods is a random and direct search method. In this method the direct search method starts from the point chosen by the user and seeks a minimum (see Siddall (1972) for a computer program). Next a new starting point is generated at random and then the direct search method seeks a better solution. The procedure is repeated t times and each time the direct search method starts from a new point where the value of t is assumed by the user. The best result from all the searches is taken as the minimum.

4.3 FURTHER POSSIBILITIES OF THE MIN–MAX APPROACH

As we discussed in Chapter 2, the min–max optimum gives a solution which treats all the criteria on terms of equal importance. At the same time the min–max approach provides further possibilities which may make the search for a satisfactory solution very efficient.

4.3.1 Weighting min–max method

Using the min–max approach together with the weighting method, a desired representation of Pareto optimal solutions can be obtained for both convex and non-convex problems. In this case the weighting coefficients are assigned to the relative deviations and then formula (2.17) has the form

$$\bigwedge_{i \in I} (z_i(\bar{x}) = \max\{w_i z_i{}'(\bar{x}), w_i z_i{}''(\bar{x})\}) \tag{4.16}$$

Since in (4.16) the weighting coefficients refer to relative deviations which are non-dimensional, the assumed values of w_i reflect exactly the priority of the criteria. We can obtain both an evenly distributed subset of Pareto optimal solutions and a subset of these solutions which is in the neighbourhood of a region of interest to the decision maker, by seeking the min–max optimum for the properly assumed values of w_i.

4.3.2 Example

Look at the results presented in Table 2.4 and illustrated graphically in Fig. 2.10. The points from 1 to 4 have been obtained using formula (4.16) for the following weighting coefficients

Points	1	2	3	4
w_1	0.2	0.4	0.6	0.8
w_2	0.8	0.6	0.4	0.2

If the decision maker wants to have more alternatives in the region between points 2 and 3, then the results obtained for the weighting coefficients as presented in Table 4.1 will give him a desirable set of solutions.

4.3.3 Displacement of ideal solution method

Another possibility in the min–max approach is contained in displacing the ideal solution. The vector f^0 has so far represented attainable minima and defined the ideal solution, which is constant for a model. However, the decision maker may want to refer to his own ideal solution which is different from the calculated one, and which reflects the goals he wants to achieve. For the assumed vector f^0 the min–max optimum will give the solution which is as close as possible to the predetermined goals. If this solution is not satisfactory two moves are possible:

(i) Choose another vector f^0 and find the min–max optimum. In this way the ideal solution is displaced closer to the preferred solution.
(ii) Choose weighting coefficients and use formula (4.16) to find the min–max optimum. In this case the weighting coefficients reflect the priority of the goals. If necessary repeat this move for new weighting coefficients.

Note that the displacement of the ideal solution enables the decision maker to concentrate on a chosen set of Pareto optimal solutions which can be gradually reduced. The concept of displacing the ideal solution can be implemented for any method in which the vector f^0 is used.

Table 4.1 Solutions in the region between points 2 and 3 for the example from Fig. 2.10

No.	w_1	w_2	$\bar{x} = [x_1, x_2]^T$	$\bar{f}(\bar{x}) = [f_1(\bar{x}), f_2(\bar{x})]^T$
1	0.45	0.55	$[236.4, 65.5]^T$	$[3.814 \cdot 10^6, 0.418 \cdot 10^{-3}]^T$
2	0.5	0.5	$[237.0, 66.4]^T$	$[3.719 \cdot 10^6, 0.424 \cdot 10^{-3}]^T$
3	0.55	0.45	$[237.2, 67.3]^T$	$[3.624 \cdot 10^6, 0.431 \cdot 10^{-3}]^T$

Table 4.2 Results of calculations for vector $\bar{f}^0 = [-900, -150]^T$ and for different weighting coefficients

No.	w_1	w_2	$\bar{x} = [x_1, x_2]^T$	$\bar{f}(\bar{x}) = [f_1(\bar{x}), f_2(\bar{x})]^T$
1	0.8	0.2	$[114.79, 75.34]^T$	$[-835.90, -114.79]^T$
2	0.6	0.4	$[122.03, 58.51]^T$	$[-780.73, -122.03]^T$
3	0.4	0.6	$[133.14, 44.75]^T$	$[-756.37, -133.14]^T$
4	0.2	0.8	$[141.24, 31.25]^T$	$[-721.25, -141.12]^T$

4.3.4 Example

Consider the production planning problem presented in Fig. 1.3. Let us assume that the decision maker is interested in finding the solution which equally satisfies the following goals:

(i) $1000 of the profit.
(ii) 200 units of product A.

Thus the assumed ideal solution is $\bar{f}^0 = [-1000, -200]^T$. For this vector $\bar{f}^0$ the min–max optimum gives the results

$$x^* = [142.8, 28.5]^T \quad \text{and} \quad f(x^*) = [-714.2, -142.8]^T.$$

The decision maker may want to displace the ideal solution choosing, for example, $f^0 = [-1000, -150]^T$ and then the optimum is

$$\bar{x}^* = [121.2, 64.6]^T \quad \text{and} \quad \bar{f}(\bar{x}^*) = [808.0, 121.2]^T.$$

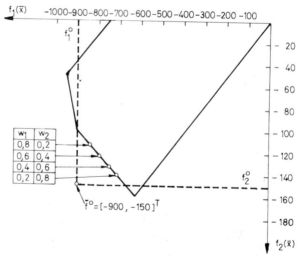

Fig. 4.7 Graphical illustration of the weighting min–max method for the given vector $\bar{f}^0$

If the decision maker wants to concentrate on the subset of Pareto optimal solutions contained between function values $f_1^0 = -900$ and $f_2^0 = -150$, then using formula (4.16) he obtains for different weighting coefficients, the solutions as presented in Table 4.2 and illustrated graphically in Fig. 4.7.

Finally, note that the possibilities of the min–max approach discussed above can be developed for interactive and iterative procedures (see the section below).

4.4 INTERACTIVE MULTICRITERION OPTIMIZATION SYSTEM

Recently, numerous multicriterion optimization methods have been oriented towards interactive on-line use. In the previous chapter we have indicated how the methods and their variants described there could be used interactively. Now we shall describe an interactive computer system which incorporates several multicriterion optimization methods useful in engineering problems. This system has been developed at the Technical University of Cracow and it is designed to:

(i) facilitate interactive processes for computer-aided decision making,
(ii) cover a wide range of decision-making problems,
(iii) provide an effective tool for seeking a preferred solution.

The system contains the following multicriterion optimization methods:

(1) Min–max method which uses (2.17) to determine the elements of the vector $\bar{z}(\bar{x})$.

(2) Global criterion method in which (3.13) is used as the global function.

(3) Weighting min–max method which uses (4.16) to determine the elements of the vector $\bar{z}(\bar{x})$.

(4) Pure weighting method in which (3.1) is used to determine a preferred solution.

(5) Normalized weighting method in which $\bar{f}(\bar{x})$ is used in (3.1).

For the methods (1), (2), (3) and (5) the ideal solution can be calculated by the system, or it can be given by the decision maker. In the latter case the ideal solution is determined by the values of the goals the decision maker wants to achieve.

The system is equipped with the following single criterion optimization methods:

(1) the flexible tolerance (FT) method,

(2) the direct and random search (DRS) method.

Other single criterion methods can be easily introduced into the system, and they are available from the author..

Note that the choice of the single criterion method may have a great influence on the result obtained.

The general flow diagram of the system is shown in Fig. 4.8. In Appendix B a FORTRAN program and a full description of the system are provided.

We shall describe the main possibilities of the system, using as the example the beam design problem presented in Fig. 1.4.

The problem is introduced into the system by means of the following subroutines:

(1) The objective functions

```
SUBROUTINE  FCELU  (X,FU)
DIMENSION  X(20),FU(20)
FU(1)=0.785*(X(1)*(6400.-X(2)*X(2))+
1    (1000.-X(1))*(10000.-X(2)*X(2)))
FU(2)=3.298E-5*((1./(4.096E7-X(2)**4)-
1    1./(1.E8-X(2)**4))*X(1)**3+1.E9/(1.E8-X(2)**4))
RETURN
END
```

(2) The inequality constraints

```
SUBROUTINE  OGRN(X,G)
DIMENSION  X(20),G(40)
G(1)=180.-9.78E6*X(1)/(4.096E7-X(2)**4)
G(2)=75.2-X(2)
G(3)=X(2)-40.
G(4)(=X(1)
RETURN
END
```

If there are equality constraints they are introduced by means of SUBROUTINE OGRR(X,H).

An example of the output list for the above problem is presented in Fig. 4.9. In this figure the data introduced by the user start with the question mark and the numbers on the right refer to the explanations given below.

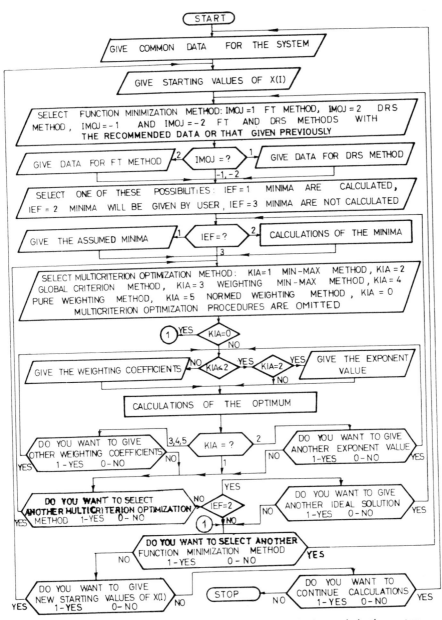

Fig. 4.8 A general flow diagram for the interactive multicriterion optimization system

```
*************************************************************************
               INTERACTIVE MULTICRITERION OPTIMIZATION SYSTEM
                     TECHNICAL UNIVERSITY OF CRACOW
                           CRACOW POLAND
          ----------------------------------------
    ---GIVE COMMON DATA FOR THE SYSTEM
 ? 2 0 4 2 0 1000. 76. 0. 40.
           COMMON DATA FOR THE SYSTEM                                    .
           NUMBER OF DECISION VARIABLES                    2            (1)
           NUMBER OF EQUALITY CONSTRAINTS                  0
           NUMBER OF INEQUALITY CONSTRAINTS               4
           NUMBER OF OBJECTIVE FUNCTIONS                   2
           INTERMEDIATE OUTPUT                             0
           ESTIMATED UPPER BOUNDS ON X(I)
                .100000E+04         .760000E+02
           ESTIMATED LOWER BOUNDS ON X(I)
                0.                 .400000E+02
    ---GIVE STARTING VALUES OF X(I)
 ? 250. 60.
           STARTING VALUES OF X(I)                                      (2)
                .250000E+03         .600000E+02
    ---SELECT FUNCTION MINIMIZATION METHOD
 ? 1
 --------------------------------------------------------------------   (3)
           YOU HAVE CHOSEN FT METHOD FOR FUNCTION MINIMIZATION
    ---GIVE DATA FOR THE METHOD
 ? .000001 10.
           DATA FOR THE METHOD
           DESIRED CONVERGENCE                        .10000E-05
           SIZE OF INITIAL POLYHEDRON                 .10000E+01
 --------------------------------------------------------------------
    ---SELECT ONE OF THESE POSSIBILITIES
       1 - MINIMA ARE CALCULATED BY THE SYSTEM
       2 - MINIMA WILL BE GIVEN BY THE USER
       3 - MINIMA ARE NOT CALCULATED
 ? 1                                                                    (4)
 --------------------------------------------------------------------
           CALCULATED MINIMA OF OBJECTIVE FUNCTIONS
 ***RESULTS OF  1  FUNCTION MINIMIZATION***
 STARTING VALUES OF X(I)
      .250000E+03         .600000E+02
 VALUE OF  1 OBJECTIVE FUNCTION  =        .2961250E+07
 VALUE OF  2 OBJECTIVE FUNCTION  =        .4976845E-03
 VECTOR OF DECISION VARIABLES
      .1590742E+03        .7520000E+02
 INEQUALITY CONSTRAINT VALUES
      .6764422E+01     -.8993361E-06         .3520000E+02     .1590742E+03
 ***RESULTS OF  2  FUNCTION MINIMIZATION***
 STARTING VALUES OF X(I)
      .250000E+03         .600000E+02
 VALUE OF  1 OBJECTIVE FUNCTION  =        .6582623E+07
 VALUE OF  2 OBJECTIVE FUNCTION  =        .3384647E-03
 VECTOR OF DECISION VARIABLES
      .4026002E+01        .4000000E+02
 INEQUALITY CONSTRAINT VALUES
      .1789746E+03        .3520000E+02     -.8786553E-06     .4026002E+01
 --------------------------------------------------------------------
 PAYOFF TABLE FOR FUNCTION INCREMENTS
 ------------------------------------
      0.                 .1592198E-03
      .3621373E+07       0.                                             (5)
 --------------------------------------------------------------------
 PAYOFF TABLE FOR FUNCTION RELATIVE INCREMENTS
 ---------------------------------------------
      0.                 .4704177E+00
      .1222920E+01       0.
    ---SELECT MULTICRITERION OPTIMIZATION METHOD
 ? 1                                                                    (6)
 --------------------------------------------------------------------
                          MIN-MAX METHOD
 --------------------------------------------------------------------
 STARTING VALUES OF X(I)
      .250000E+03         .600000E+02
 VALUE OF THE MINIMIZED QUANTITY  =       .2545275E+00
 VALUE OF  1 OBJECTIVE FUNCTION  =       .3714969E+07
 VALUE OF  2 OBJECTIVE FUNCTION  =       .4246133E-03
 VECTOR OF DECISION VARIABLES
      .2370261E+03        .6643991E+02
 INEQUALITY CONSTRAINT VALUES
      .7205164E+02        .8760094E+01     .2643991E+02     .2370261E+03
```

```
---DO YOU WANT TO SELECT ANOTHER MULTICRITERION OPTIMIZATION METHOD
? 1
---SELECT MULTICRITERION OPTIMIZATION METHOD
? 2
-----------------------------------------------------------------------
                         GLOBAL CRITERION METHOD
-----------------------------------------------------------------------
---GIVE THE EXPONENT VALUE
? 2
SOLUTION  1
      EXPONENT =    2
STARTING VALUES OF X(I)
      .250000E+03    .600000E+02
VALUE OF THE MINIMIZED QUANTITY =       .1231585E+00
VALUE OF  1 OBJECTIVE FUNCTION =       .3537827E+07
VALUE OF  2 OBJECTIVE FUNCTION =       .4372869E-03
VECTOR OF DECISION VARIABLES
      .2371542E+03    .6811357E+02
INEQUALITY CONSTRAINT VALUES
      .6066285E+02    .7086428E+01    .2811357E+02    .2371542E+03
---DO YOU WANT TO GIVE ANOTHER EXPONENT VALUE
? 0
---DO YOU WANT TO SELECT ANOTHER MULTICRITERION OPTIMIZATION METHOD
? 1
---SELECT MULTICRITERION OPTIMIZATION METHOD
? 3
-----------------------------------------------------------------------
                     WEIGHTING MIN-MAX METHOD
-----------------------------------------------------------------------
---GIVE THE WEIGHTING COEFFICIENTS
? .2 .8
SOLUTION  1
      DELTA( 1)=  .20
      DELTA( 2)=  .80
STARTING VALUES OF X(I)
      .250000E+03    .600000E+02
VALUE OF THE MINIMIZED QUANTITY =       .1047330E+00
VALUE OF  1 OBJECTIVE FUNCTION =       .4511953E+07
VALUE OF  2 OBJECTIVE FUNCTION =       .3827753E-03
VECTOR OF DECISION VARIABLES
      .2247285E+03    .5867935E+02
INEQUALITY CONSTRAINT VALUES
      .1044829E+03    .1652065E+02    .1867935E+02    .2247285E+03
---DO YOU WANT TO GIVE OTHER WEIGHTING COEFFICIENTS
? 1
---GIVE THE WEIGHTING COEFFICIENTS
? .4 .6
SOLUTION  2
      DELTA( 1)=  .40
      DELTA( 2)=  .60
STARTING VALUES OF X(I)
      .224728E+03    .586794E+02
VALUE OF THE MINIMIZED QUANTITY =       .1297398E+00
VALUE OF  1 OBJECTIVE FUNCTION =       .3921729E+07
VALUE OF  2 OBJECTIVE FUNCTION =       .4116519E-03
VECTOR OF DECISION VARIABLES
      .2354730E+03    .6447064E+02
INEQUALITY CONSTRAINT VALUES
      .8276371E+02    .1072936E+02    .2447064E+02    .2354730E+03
---DO YOU WANT TO GIVE OTHER WEIGHTING COEFFICIENTS
? 0
---DO YOU WANT TO SELECT ANOTHER MULTICRITERION OPTIMIZATION METHOD
? 1
---SELECT MULTICRITERION OPTIMIZATION METHOD
? 4
-----------------------------------------------------------------------
                         PURE WEIGHTING METHOD
-----------------------------------------------------------------------
---GIVE THE WEIGHTING COEFFICIENTS
? .2 .8
SOLUTION  1
      LAMBDA( 1)=  .20
      LAMBDA( 2)=  .80
STARTING VALUES OF X(I)
      .250000E+03    .600000E+02
VALUE OF THE MINIMIZED QUANTITY =       .5922500E+06
VALUE OF  1 OBJICTIVE FUNCTION =       .2961250E+07
VALUE OF  2 OBJICTIVE FUNCTION =       .4976845E-03
VECTOR OF DECISION VARIABLES
      .1590742E+03    .7520000E+02
INEQUALITY CONSTRAINT VALUES
      .6764422E+01   -.3993361E-06    .3520000E+02    .1590742E+03
---DO YOU WANT TO GIVE OTHER WEIGHTING COEFFICIENTS
? 0
```

(7)

(8)

```
  ---DO YOU WANT TO SELECT ANOTHER MULTICRITERION OPTIMIZATION METHOD
? 1
  ---SELECT MULTICRITERION OPTIMIZATION METHOD
? 5
------------------------------------------------------------------------
                         NORMALIZED WEIGHTING METHOD
------------------------------------------------------------------------
  ---GIVE THE WEIGHTING COEFFICIENTS
? .2 .8
  SOLUTION  1
          LAMBDA( 1)=  .20
          LAMBDA( 2)=  .80
  STARTING VALUES OF X(I)
        .250000E+03          .600000E+02
  VALUE OF THE MINIMIZED QUANTITY =       .1201389E+01
  VALUE OF  1 OBJICTIVE FUNCTION =        .5110397E+07
  VALUE OF  2 OBJICTIVE FUNCTION =        .3622578E-03
  VECTOR OF DECISION VARIABLES
        .2066792E+03          .5240129E+02
  INEQUALITY CONSTRAINT VALUES
        .1195177E+03          .2279871E+02          .1240129E+02        .2066792E+03
  ---DO YOU WANT TO GIVE OTHER WEIGHTING COEFFICIENTS
? 0
  ---DO YOU WANT TO SELECT ANOTHER MULTICRITERION OPTIMIZATION METHOD
? 0
  ---DO YOU WANT TO SELECT ANOTHER FUNCTION MINIMIZATION METHOD
? 1
  ---SELECT FUNCTION MINIMIZATION METHOD
? 2
------------------------------------------------------------------------
          YOU HAVE CHOSEN DRS METHOD FOR FUNCTION MINIMIZATION
  ---GIVE DATA FOR THE METHOD
? 300 300 .01 .0001 500M 3 3
          DATA FOR THE METHOD
  MAXIMUM NUMBER OF MOVES PERMITTED                                    300
  NUMBER OF TEST POINTS IN SHOTGUN SEARCH                              300
  FRACTION OF RANGE USED AS STEP SIZE                             .10000E-01
  STEP SIZE FRACTION USED AS CONVERGENCE CRITERION               .10000E-03
  NUMBER OF RANDOMLY GENERATED POINTS FOR FINDING
  A NEW STARTING POINT                                                 500
  NUMBER OF DIRECT SEARCH METHOD RUNS USING NEW
  STARTING POINTS                                                       3
  NUMBER OF SHOTGUN SEARCHES PERMITTED                                  3
------------------------------------------------------------------------
  ---SELECT ONE OF THESE POSSIBILITIES
  1 - MINIMA ARE CALCULATED BY THE SYSTEM
  2 - MINIMA WILL BE GIVEN BY THE USER
  3 - MINIMA ARE NOT CALCULATED
? 1
------------------------------------------------------------------------
          CALCULATED MINIMA OF OBJECTIVE FUNCTIONS
------------------------------------------------------------------------
  ***RESULTS OF  1  FUNCTION MINIMIZATION***
  STARTING VALUES OF X(I)
        .250000E+03          .600000E+02
  VALUE OF  1 OBJECTIVE FUNCTION =        .2944492E+07
  VALUE OF  2 OBJECTIVE FUNCTION =        .4990588E-03
  VECTOR OF DECISION VARIABLES
        .1661247E+03          .7517317E+02
  INEQUALITY CONSTRAINT VALUES
        .5201706E-03          .2682853E-01          .3517317E+02        .1661247E+03
  ***RESULTS OF  2  FUNCTION MINIMIZATION***
  STARTING VALUES OF X(I)
        .250000E+03          .600000E+02
  VALUE OF  1 OBJECTIVE FUNCTION =        .6593969E+07
  VALUE OF  2 OBJECTIVE FUNCTION =        .3384647E-03
  VECTOR OF DECISION VARIABLES
        .1098633E-01          .4000000E+02
  INEQUALITY CONSTRAINT VALUES
        .1799972E+03          .3520000E+02          .4882799E-05        .1098633E-01
------------------------------------------------------------------------
  PAYOFF TABLE FOR FUNCTION INCREMENTS
------------------------------------------------------------------------
    0.              .1605941E-03
      .3649477E+07  0.
------------------------------------------------------------------------
  PAYOFF TABLE FOR FUNCTION RELATIVE INCREMENTS
------------------------------------------------------------------------
    0.              .4744781E+00
      .1239425E+01  0.
  ---SELECT MULTICRITERION OPTIMIZATION METHOD
? 0
  ---DO YOU WANT TO SELECT ANOTHER FUNCTION MINIMIZATION METHOD
? 1
```

```
---SELECT FUNCTION MINIMIZATION METHOD
? -1
---SELECT ONE OF THESE POSSIBILITIES
    1 - MINIMA ARE CALCULATED BY THE SYSTEM
    2 - MINIMA WILL BE GIVEN BY THE USER
    3 - MINIMA ARE NOT CALCULATED
? 2
---GIVE THE ASSUMED MINIMA
? .3E7 .4E-3
----------------------------------------------------------------
        ASSUMED MINIMA OF OBJECTIVE FUNCTIONS
----------------------------------------------------------------
        MINIMUM OF  1  OBJECTIVE FUNCTION =   .30000E+07
        MINIMUM OF  2  OBJECTIVE FUNCTION =   .40000E-03
---SELECT MULTICRITERION OPTIMIZATION METHOD
? 1
----------------------------------------------------------------
                    MIN-MAX METHOD
----------------------------------------------------------------
STARTING VALUES OF X(I)
    .250000E+03      .600000E+02
VALUE OF THE MINIMIZED QUANTITY =      .1257587E+00
VALUE OF  1 OBJECTIVE FUNCTION =    .3377276E+07
VALUE OF  2 OBJECTIVE FUNCTION =    .4503035E-03
VECTOR OF DECISION VARIABLES
    .2361121E+03      .6962567E+02
INEQUALITY CONSTRAINT VALUES
    .4774080E+02      .5574331E+01      .2962567E+02      .2361121E+03
---DO YOU WANT TO SELECT ANOTHER MULTICRITERION OPTIMIZATION METHOD
? 10
---DO YOU WANT TO SELECT ANOTHER IDEAL SOLUTION
? 1
---GIVE THE ASSUMED MINIMA
? .31E7 .44E-3
----------------------------------------------------------------
        ASSUMED MINIMA OF OBJECTIVE FUNCTIONS
----------------------------------------------------------------
        MINIMUM OF  1  OBJECTIVE FUNCTION =   .31000E+07
        MINIMUM OF  2  OBJECTIVE FUNCTION =   .44000E-03
---SELECT MULTICRITERION OPTIMIZATION METHOD
? 1
----------------------------------------------------------------
                    MIN-MAX METHOD
----------------------------------------------------------------
STARTING VALUES OF X(I)
    .250000E+03      .600000E+02
VALUE OF THE MINIMIZED QUANTITY =      .4919333E-01
VALUE OF  1 OBJECTIVE FUNCTION =    .3252499E+07
VALUE OF  2 OBJECTIVE FUNCTION =    .4616451E-03
VECTOR OF DECISION VARIABLES
    .2344021E+03      .7080142E+02
INEQUALITY CONSTRAINT VALUES
    .3519605E+02      .4398580E+01      .3080142E+02      .2344021E+03
---DO YOU WANT TO SELECT ANOTHER MULTICRITERION OPTIMIZATION METHOD
? 0
---DO YOU WANT TO SELECT ANOTHER IDEAL SOLUTION
? 0
---DO YOU WANT TO SELECT ANOTHER FUNCTION MINIMIZATION METHOD
? 0
---DO YOU WANT TO GIVE NEW STARTING VALUES OF X(I)
? 1
---GIVE STARTING VALUES OF X(I)
? 222. 65.
        STARTING VALUES OF X(I)
            .222000E+03      .650000E+02
---SELECT FUNCTION MINIMIZATION METHOD
? -2
---SELECT ONE OF THESE POSSIBILITIES
    1 - MINIMA ARE CALCULATED BY THE SYSTEM
    2 - MINIMA WILL BE GIVEN BY THE USER
    3 - MINIMA ARE NOT CALCULATED
? 1
----------------------------------------------------------------
        CALCULATED MINIMA OF OBJECTIVE FUNCTIONS
----------------------------------------------------------------
***RESULTS OF  1  FUNCTION MINIMIZATION***
STARTING VALUES OF X(I)
    .222000E+03      .650000E+02
VALUE OF  1 OBJECTIVE FUNCTION =    .2945539E+07
VALUE OF  2 OBJECTIVE FUNCTION =    .4988133E-03
VECTOR OF DECISION VARIABLES
    .1672276E+03      .7513789E+02
INEQUALITY CONSTRAINT VALUES
    .3160247E-03      .6211192E-01      .3513789E+02      .1672276E+03
```

(9)

(10)

```
***RESULTS OF  2  FUNCTION MINIMIZATION***
STARTING VALUES OF X(I)
   .222000E+03      .650000E+02
VALUE OF  1 OBJECTIVE FUNCTION =    .6593973E+07
VALUE OF  2 OBJECTIVE FUNCTION =    .3384647E-03
VECTOR OF DECISION VARIABLES
   .9033203E-02      .4000002E+02
INEQUALITY CONSTRAINT VALUES
   .1799977E+03      .3519998E+02      .1708982E-04      .9033203E-02
------------------------------------------
PAYOFF TABLE FOR FUNCTION INCREMENTS
------------------------------------------
   0.                .1603486E-03
   .3648435E+07      0.
------------------------------------------
PAYOFF TABLE FOR FUNCTION RELATIVE INCREMENTS
------------------------------------------
   0.                .4737528E+00
   .1238631E+01      0.
   ---SELECT MULTICRITERION OPTIMIZATION METHOD
? 0
   ---DO YOU WANT TO SELECT ANOTHER FUNCTION MINIMIZATION METHOD
? 0
   ---DO YOU WANT TO GIVE NEW STARTING VALUES OF X(I)
? 0
   ---DO YOU WANT TO CONTINUE CALCULATIONS
? 1
   ---GIVE COMMON DATA FOR THE SYSTEM
? 2 0 4 1 0 1000. 80. 0. 40                                              (11)
           COMMON DATA FOR THE SYSTEM
           NUMBER OF DECISION VARIABLES                 2
           NUMBER OF EQUALITY CONSTRAINTS               0
           NUMBER OF INEQUALITY CONSTRAINTS             4
           NUMBER OF OBJECTIVE FUNCTIONS                1
           INTERMEDIATE OUTPUT                          0
           ESTIMATED UPPER BOUNDS ON X(I)
              .100000E+04      .800000E+02
           ESTIMATED LOWER BOUNDS ON X(I)
              0.               .400000E+02
   ---GIVE STARTING VALUES OF X(I)
? 200. 65.
           STARTING VALUES OF X(I)
              .200000E+03      .650000E+02
   ---SELECT FUNCTION MINIMIZATION METHOD
? -1
   ---SELECT ONE OF THESE POSSIBILITIES
     1 - MINIMA ARE CALCULATED BY THE SYSTEM
     2 - MINIMA WILL BE GIVEN BY THE USER
     3 - MINIMA ARE NOT CALCULATED
? 1
------------------------------------------------------------------------
           CALCULATED MINIMA OF OBJECTIVE FUNCTIONS
------------------------------------------------------------------------
***RESULTS OF  1  FUNCTION MINIMIZATION***
STARTING VALUES OF X(I)
   .200000E+03      .650000E+02
VALUE OF  1 OBJECTIVE FUNCTION =    .2972835E+07
VECTOR OF DECISION VARIABLES
   .1549746E+03      .7520000E+02
INEQUALITY CONSTRAINT VALUES
   .1122902E+02     -.1139609E-05      .3520000E+02      .1549746E+03
   ---SELECT MULTICRITERION OPTIMIZATION METHOD
? 0
   ---DO YOU WANT TO SELECT ANOTHER FUNCTION MINIMIZATION METHOD
? 0
   ---DO YOU WANT TO GIVE NEW STARTING VALUES OF X(I)
? 0
   ---DO YOU WANT TO CONTINUE CALCULATIONS
? 0
   ---END OF JOB,THANK YOU---
SRU    28.344 UNTS.                                                      (12)
RUN COMPLETE.
```

Fig. 4.9 An example of output list for the multicriterion optimization system

(1) The data introduced by the user are printed with their description. This allows the user to check the correctness of the data.

Note that the estimated upper and lower bounds on x_i are not treated as constraints, but are used to generate a feasible starting point.

(2) Starting values of x_i should give a feasible point. If the user introduces a non-feasible point the program will try to find its own feasible point. Note that the results we obtain using the FT and DRS methods may depend significantly on the starting point and this is normal for most minimization methods.

(3) Two function minimization methods are introduced, and thus we may select

1   FT method

2   DRS method

If we put a minus sign before these numbers, the program will use the data for the FT and DRS methods from the previous calculations, or for the first run of this part of the program the generally recommended data for both methods will be used.

(4) The following selection means are provided:

1   The program will find the ideal solution, i.e., the separately attainable minima of the objective functions.

2   The ideal solution is given by the user and then this solution defines the minima he wants to achieve. Note that he may displace this solution (see point 9).

3   This case has two meanings: (i) the minima are not used in the pure weighting method, and (ii) we do not want to change the previously defined (calculated or given) ideal solution. In other words we want to retain the ideal solution from the previous calculation.

(5) For the ideal solution found by the system the pay-off tables for the function increments and the function relative increments are calculated and printed.

(6) The system is designed to solve multicriterion optimization problems using five different methods. Selecting the numbers from 1 to 5 we choose one of these methods according to the description given in Fig. 4.8. Moreover we may omit the multicriterion optimization procedure if we select 0 at this step.

(7) For the multicriterion optimization methods which use different data the solutions are numbered $1, 2, \ldots$, within each method.

(8) For the second and the following run of the program for each multicriterion method, the program assumes that the starting point is the solution obtained in the previous run. This makes the search more efficient.

(9) The user may displace the ideal solution by giving other minima which would reflect his goals.

(10) The user may introduce other starting values of x_i.

(11) The user may change the number of objective functions and constraints which are considered by the system. Thus it is possible to change the optimization model by reducing and/or increasing the values of k, m and p. If we want to do this, we have to prepare the subroutines FCELU, OGRN and OGRR as appropriate. In the listing, it is shown how the system responds while solving a single criterion optimization problem.

(12) This example has been run on a Cyber 72 computer.

The interactive computer system described above appears to be a very effective tool in solving optimization problems the engineer may meet. The system has been applied to solve real-world multicriterion optimization problems and in each case the preferred solution has been found in a short computation time. Most examples from this book have been calculated using the system.

CHAPTER 5

Network multicriterion optimization

Many engineering systems may be considered as networks, for example transportation, transmission and communication systems. The common structure of these systems is a collection of points, usually called nodes, and a collection of lines, usually called arcs, which connect these nodes. If the connection between two nodes is directed the networks of this type are called directed networks. If only one value, for example cost, is associated with an arc then the problem is of the single criterion type. Problems of this type have a rich bibliography and several books can be recommended: Christofides (1975), Hu (1970), Mandl (1979) and Steenbrink (1974). The problem becomes much more complicated if we want to associate several values with an arc, for example cost and time. In this case we have to deal with the multicriterion optimization problem. Although multicriterion network modelling has a wide engineering application, little research has been done in this field and only few papers can be quoted: Hansen (1980), Marchet and Siskos (1979), Osyczka (1980) and Vincke (1974).

In this chapter multicriterion problems arising within a class of directed and acyclic networks will be discussed. Such networks are most commonly used for engineering problems.

The problem is formulated as follows: find the path in the network for which the sums of values which represent the objective functions are optimal. For this problem the Pareto and min–max optimal paths are defined and a method of seeking the preferred path is described.

5.1 PROBLEM FORMULATION

Let $S = \langle U, R, \bar{f} \rangle$ denote a network whose graph $G = \langle U, R \rangle$ belongs to the class of directed and acyclic graphs, where U is a set of nodes, R a two argument relation defined on the set U, and $\bar{f}$ a vector function defined on the arcs of the graph G. The two argument relation R defines which nodes are connected by arcs and in which direction. For example, $u_r R u_s$ denotes that the node u_r is connected with the node u_s and that the direction is from

u_r to u_s. The vector function $\bar{f}$ is defined as follows. A vector

$$\bar{f}(u_r, u_s) = [f_1(u_r, u_s), ..., f_i(u_r, u_s), ..., f_k(u_r, u_s)]^T$$

is associated with any pair of nodes $u_r, u_s \in U$ for which $u_r R u_s$. The k components of the vector $\bar{f}(u_r, u_s)$ which are given as the weights assigned to the arc connecting the pair of nodes u_r, u_s, represent non-commensurable criteria which we must consider. Note that the term 'weight' has here a different meaning to that used in the weighting objective method. The term 'length' is also used instead of 'weight'.

We denote

Γ_{u_s} $-$ a set of the graph nodes $u_r \in U$ for which $u_s R u_r$

$\Gamma_{u_s}^{-1}$ $-$ a set of the graph nodes $u_r \in U$ for which $u_r R u_s$

In other words the set Γ_{u_s} contains those nodes for which the arcs lead out of the node u_s whereas, the set $\Gamma_{u_s}^{-1}$ contains those nodes for which the arcs lead into the node u_s.

We assume that:

(1) The graph G has two privileged nodes $-$ the initial node such that

$\Gamma_{u_1}^{-1} = \emptyset$ and the terminal node such that $\Gamma_{u_n} = \emptyset$.

(2) For any node $u_r \notin \{u_1, u_n\}$ in the graph G we have $\Gamma_{u_r} \neq \emptyset$ and $\Gamma_{u_r}^{-1} \neq \emptyset$. This means that any node u_r which is neither the initial node nor the terminal node, has at least one arc which leads into this node and at least one which leads out of this node.

(3) Each component of the vector $F(u_r, u_s)$ can only assume non-negative integer values, i.e.,

$$\bigwedge_{u_r, u \in U} \bigwedge_{i \in I} (f_i(u_r, u_s) \geqslant 0) \tag{5.1}$$

We denote

$d_j = \{u_1, ..., u_r, ..., u_n\}$ the jth path in the graph joining the initial node u_1 to the terminal node u_n,

$D = \{d_j\}$ a set of all the d_j paths in the graph, where $j = \{1, 2, ..., J\}$.

The problem is formulated as follows.

Find a path $d^* = \{u_1^*, ..., u_r^*, ..., u_n^*\}$ in the network $S = \langle U, R, \bar{f} \rangle$ which optimizes a vector function $\bar{f}(d_j) = [f_1(d_j), ..., f_i(d_j), ..., f_k(d_j)]^T$ defined in the k-dimensional space E^k. The ith component of the vector $f(d_j)$ is determined from

$$f_i(d_j) = \pm \sum_{u_r, u_s \in d_j} f_i(u_r, u_s) \tag{5.2}$$

where sign ' $+$ ' refers to the functions which are to be minimized, and sign ' $-$ ' to the functions which are to be maximized.

Note that because of the assumption (5.1) the functions which are to be maximized can be transformed to a form that permits minimization after the sum of $f_i(u_r, u_s)$ has been calculated.

The ideal vector $f^0 = [f_1^0, ..., f_i^0, ..., f_k^0]^T$ can be determined after finding the extreme paths for each criterion separately. Let $d^{0(i)} = \{u_1^{0(i)}, ..., u_r^{0(i)}, ..., u_n^{0(i)}\}$ be the path which minimizes or maximizes the ith objective function $f_i(d_j)$. In other words the path $d^{0(i)} \in D$ is such that for the objective functions which are to be minimized we have

$$f_i^0 = f_i(d^{0(i)}) = \min_{d_j \in D} \sum_{u_r, u_s \in d_j} f_i(u_r, u_s) \tag{5.3}$$

and for the objective functions which are to be maximized we have

$$f_i^0 = f_i(d^{0(i)}) = \max_{d_j \in D} \sum_{u_r, u_s \in d_j} f_i(u_r, u_s) \tag{5.4}$$

In network terminology the path which satisfies (5.3) is called the shortest path and at which satisfies (5.4) the longest path. Many algorithms have been developed to find these two kinds of paths in the network.

Example 5.1

Suppose that we have to travel by car between two places in a town with heavy traffic. We may choose different streets to get to the place of destination. We may draw the graph $G = \langle U, R \rangle$ which illustrates all possible routes. Let the graph G shown in Fig. 5.1 represent these possibilities. The nodes denote different places in the town and the arcs the streets between these places. The node u_1 is the starting point of the journey, and the node u_8 is the place of destination. Let us assume that we want to travel from x_1 to x_8 at the lowest cost. It is usually impossible to estimate the cost of the parts of the routes, but we may estimate the time and the distance. These two values are not proportionally related because of traffic lights, pedestrian ways and so on, but together approximate in some way the cost of the journey. Of course, they cannot be added together, but they can be treated

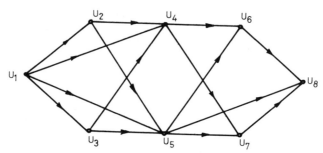

Fig. 5.1 Graph $G = \langle U, R \rangle$ for Example 5.1

as two criteria in the optimization model. Assuming that we can estimate the mean traverse time at a given time of the day and the distance for each arc of the graph G, we can create the network $S = \langle U, R, f \rangle$ presented in Fig. 5.2. In this network the time is given in minutes and the distance in kilometres.

Now the problem is to find in the network S such a path d^* for which both criteria, i.e., the time and the distance are minimal.

The ideal vector f^0 can be determined after finding the shortest paths for each criterion separately. For this network we have

(1) For the time criterion

$$d^{0(1)} = \{u_1, u_5, u_8\}, f_1^0 = 45 \text{ min}$$

(2) For the distance criterion

$$d^{0(2)} = \{u_1, u_2, u_4, u_6, u_8\}, f_2^0 = 22 \text{ km}$$

and thus $\bar{f}^0 = [45, 22]^T$

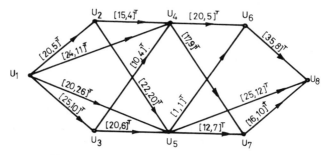

Fig. 5.2 Network $S = \langle U, R, f \rangle$ for Example 5.1

Note that the so-called tree networks can be transformed to the class of networks considered in this chapter by adding a node and connecting this node with all branches of the tree by means of additional arcs which are weighted by zero vectors. The tree graphs and networks represent typical decision-making situations.

Example 5.2

Suppose that we have to travel from the town A to town B and there is no through train, bus or plane connection between these towns. The nearest town to which there is a through train or bus connection is the town C. The plane connection is to the town D. All possible ways of travelling may be thus modelled by the graph presented in Fig. 5.3 where the arcs mean

u_1Ru_2 – own car to the town B

u_1Ru_3 – train from A to C

u_1Ru_4 – bus from A to C

u_1Ru_5 – plane from A to D

u_3Ru_6 and u_4Ru_9 – train from C to B

u_3Ru_7 and u_4Ru_{10} – train from C to B

u_3Ru_8 and u_4Ru_{11} – rent a car and drive from C to B

u_5Ru_{12} – train from D to B

u_5Ru_{13} – bus from D to B

u_5Ru_{14} – rent a car and drive from D to B

Let us assume that we want to get to the town B having in mind two criteria, the time and cost. We may easily estimate these values for each arc of the graph and create the network as presented in Fig. 5.4. The time of waiting for the connection can be added to the time of travelling. In the network

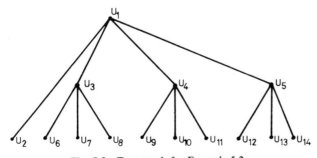

Fig. 5.3 Tree graph for Example 5.2

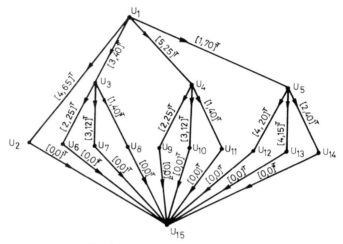

Fig. 5.4 Network for Example 5.2

presented in Fig. 5.4 the time is given in hours, the cost in dollars. Now the problem of finding a preferred way of travelling is to find a preferred path in the network S.

5.2 PARETO AND MIN–MAX OPTIMAL PATHS

5.2.1 *Pareto optimal path*

A path $d^* \in D$ is Pareto optimal if for every $d_j \in D$ either

$$\bigwedge_{i \in I} (f_i(d_j) = f_i(d^*))$$ (5.5)

or, there is at least one $i \in I$ such that

$$f_i(d_j) > f_i(d^*)$$ (5.6)

Of course, usually we shall have a set of Pareto optimal paths non-inferior paths. We use D^p to denote this set and F^p to denote the map D^p in the space of objectives E^k.

For the network presented in Fig. 5.2 we have the following set of Pareto optimal paths

$$D^p = \{d_1, d_2, d_3, d_4, d_5, d_6, d_7\}$$

where $d_1 = \{u_1, u_2, u_4, u_6, u_8\}, \quad \bar{f}(d_1) = [90, 22]^T$

$d_2 = \{u_1, u_4, u_6, u_8\}, \qquad \bar{f}(d_2) = [79, 24]^T$

$d_3 = \{u_1, u_3, u_5, u_8\}, \qquad \bar{f}(d_3) = [70, 28]^T$

$d_4 = \{u_1, u_2, u_4, u_7, u_8\}, \quad \bar{f}(d_4) = [68, 28]^T$

$d_5 = \{u_1, u_4, u_7, u_8\}, \qquad \bar{f}(d_5) = [57, 30]^T$

$d_6 = \{u_1, u_5, u_6, u_8\}, \qquad \bar{f}(d_6) = [56, 35]^T$

$d_7 = \{u_1, u_5, u_8\}, \qquad\quad \bar{f}(d_7) = [45, 38]^T$

In the space of objectives these paths give the points denoted by heavy dots in Fig. 5.5. The light dots refer to the remaining paths in the network.

5.2.2 Min–max optimal path

Let $\bar{z}(d_j) = [z_1(d_j), ..., z_i(d_j), ...,z_k(d_j)]^T$ be a vector of relative increments of the objective functions for the path d_j. The components of the vector $\bar{z}(d_j)$ are defined as follows

$$\bigwedge_{i \in I} z_i(d_j) = \max\{z_i{}'(d_j), z_i{}''(d_j)\} \tag{5.7}$$

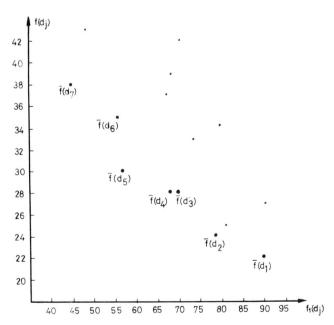

Fig. 5.5 Graphical representation of the Pareto optimal paths

where

$$z_i'(d_j) = \frac{|f_i(d_j) - f_i^0|}{|f_i^0|} \tag{5.8}$$

$$z_i''(d_j) = \frac{|f_i(d_j) - f_i^0|}{|f_i(d_j)|} \tag{5.9}$$

A path $d^* \in D$ is optimal in the min–max sense if for every $d_j \in D$ the following recurrence formula is satisfied:

Step 1

$$v_1(d^*) = \min_{d_j \in D} \max_{i \in I} \{z_i(d_j)\}$$

and then $I_1 = \{i_1\}$, where i_1 is the index for which the value of $z_i(d_j)$ is maximal

If there is a set of paths $D_1 \subset D$ which satisfies Step 1, then

Step 2

$$v_2(d^*) = \min_{d_j \in D} \max_{\substack{i \in I \\ i \notin I^1}} \{z_i(d_j)\}$$

and then $I_2 = \{i_1, i_2\}$, where i_2 is the index for which the value of $z_i(d_j)$ in this step is maximal.

. . . . (5.10)

If there is a set of paths $D_{r-1} \in D$ which satisfies Step $k - 1$, then

Step r

$$v_r(d^*) = \min_{d_j \in D} \max_{\substack{i \in I \\ i \notin I_{r-2}}} \{z_i(d_j)\}$$

and then $I_r = \{I_{r-1}, i_r\}$, where i_r is the index for which the value of $z_i(d_j)$ in the rth step is maximal.

Step k

$$v_k(d^*) = \min_{d_j \in D_{k-1}} \{z_i(d_j)\} \qquad \text{for } i \in I \text{ and } i \notin I_{k-1}$$

where $v_1(d^*), ..., v_k(d^*)$ is the set of optimal values of relative increments ordered non-increasingly.

For the network presented in Fig. 5.2 the optimal path in the min–max sense is

$$d^* = \{u_1, u_4, u_7, u_8\}, \qquad \bar{f}(d^*) = [57, 30]^T, \qquad \bar{z}(d^*) = [0.266, 0.363]^T$$

This path has been determined by the first step of formula (5.10). For the network presented in Fig. 5.4 the two following paths satisfy the first step of formula (5.10).

$$d_1 = \{u_1, u_2, u_{15}\}, \qquad f(d_1) = [4, 65]^T, \qquad z(d_1) = [0.333, 0.756]^T$$

$$d_2 = \{u_1, u_3, u_6, u_{15}\}, \qquad f(d_2) = [5, 65]^T, \qquad z(d_2) = [0.666, 0.756]^T$$

The second step of (5.10) selects the path d_1 as the optimal path in the min–max sense.

Obviously for the network optimization the decision-making problem is the same as for mathematical programming (see Section 2.4), but the methods of seeking the preferred path are less developed.

5.3 *Method of solution*

In Section 4.1 we described two algorithms which enable us to select the set of Pareto optimal solutions and the solution optimal in the min–max sense from any given set of feasible solutions. For the network optimization problem, the set of feasible solutions is represented by the set of all paths in the network. Assuming that we may examine all paths from the set D and applying the PARETO or MINMAX algorithm we may select the set D^p or the path d^*. For moderate sized networks we may examine all the paths $d_j \in D$ in a reasonable computational time but generally the problem is intractable, since the number of paths to be examined grows exponentially with the size of network.

We shall describe a more efficient method of solution which is based on the Osyczka (1980) approach. The essence of the method consists in the ordered examination of the paths and this process is terminated if a defined condition which will be discussed later, has been satisfied. This ordered examination of the paths consists in seeking suboptimal solutions in the network for the arbitrarily chosen objective function. It means that the first path is the minimum sum path, i.e., the shortest path. Each subsequent path differs from the previous path and gives an equal or greater value of the objective function than the previous one.

We assume that index $i = 1$ is assigned to the objective function for which the suboptimal paths are to be determined. Hence, the problem of finding the sequence of suboptimal paths can be formulated as follows.

Find the path $d^{(l)} = \{u_1^{(l)}, \ldots, u_n^{(l)}\}$ such that for $l = 1$

$$f_1(d_1) = \min_{\substack{d^{(l)} \in D}} \sum_{u_r, s \in d^{(l)}} f_1(u_r, u_s) \tag{5.11}$$

and for $l > 1$

$$f_1(d^{(l)}) \geqslant f_1(d^{(l-1)}) \quad \text{and} \quad d^{(l)} \neq d^{(l-1)} \tag{5.12}$$

An algorithm for solving the above problem has been presented by Osyczka (1976). Obviously, if we want to apply this algorithm to our method we have to assume that at least one objective fucntion is to be minimized. Otherwise Maas-Teugels (1969) algorithm has to be applied. This algorithm solves the problem of seeking the suboptimal paths in the network for the objective function which is to be maximized.

Let d_j^p denote the jth Pareto optimal path and $f_j^p = [f_1(d_j^p), \ldots, f_k(d_j^p)]^T$ a vector of the objective functions for the path d_j^p. Now the procedure of seeking the set of Pareto optimal paths contained in the selected region is as follows

    (1) Read the network $S = \langle U, R, f \rangle$.
    (2) Set $f_1^p = \infty$ for $i = 1, 2, \ldots, k$, $v_1^* = \infty$, $l^* = 0$, $j^a = 1$.
    (3) Find the extreme paths for each criterion separately, i.e., find $d^{0(i)}$ and f_i^0 for $i = 1, 2, \ldots, k$.
    (4) Set $l = 1$.
    (5) Find the path $d^{(l)}$ which satisfies (5.11) and (5.12).
    (6) Determine values of the objective functions for the path $d^{(l)}$, i.e., determine $f_i(d^{(l)})$ for $i = 1, 2, \ldots, k$.
    (7) Call subroutine PARETO in order to check if the path $d^{(l)}$ should be stored as the Pareto optimal.
    (8) Call subroutine MINMAX in order to check if the path $d^{(l)}$ is the min–max optimal.
    (9) If

$$v_1^* < z_1(d^{(l)}) \tag{5.13}$$

then $l^* = l$ and go to 10; otherwise go straight to 10.
(10) Set $l = l + 1$
(11) If $l \leqslant l^a$ then go to 5; otherwise go to 12
(12) Print $d_j^p f_j^p$ for $j = 1, 2, \ldots, j^a$, d^*, $\bar{f}(d^*)$, l^*.

The general flow-diagram of this procedure is presented in Fig. 5.6. In this figure we use the MINMAX and PARETO subroutines described in Section 4.1. Obviously, using these subroutines here we have to modify

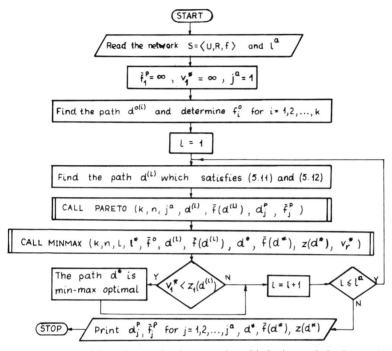

Fig. 5.6 A general flow diagram for the network multicriterion optimization method

them replacing the vector x by the path d. The FORTRAN program with the modified subroutines MINMAX and PARETO is given in Appendix C.

Two things need more detailed explanation. The first refers to the set of Pareto optimal paths. The number of suboptimal paths we want to consider is assumed *a priori*. Thus not all the Pareto optimal paths will be found but only those which are contained in the region defined by the cone

$$C_1^{l^a} = (C^- + \bar{f}_1^{l^a})$$

where $\bar{f}_1^{l^a} = [f_1(d^{(l^a)}), f_2(d^{0(1)}), f_3(d^{0(1)}), ..., f_k(d^{0(1)})]^T$.

For Example 5.1, the graphical illustration of the cone C_1^7 for $l^a = 7$ is given in Fig. 5.7.

If none of the Pareto optimal paths from the cone C^1 satisfies the decision maker he may repeat the above procedure for

(i) a greater value of l^a,

(ii) another objective function for which the suboptimal paths are determined.

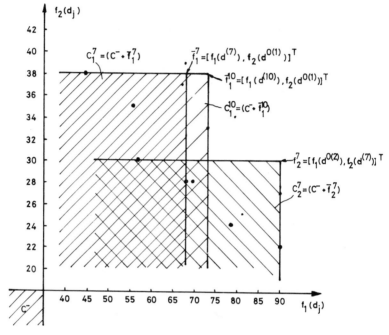

Fig. 5.7 Graphical illustration of cones containing the Pareto optimal paths

For Example 5.1, if we increase l^a from 7 to 10 we obtain the set of Pareto optimal paths contained in the cone C_1^{10}, whereas if we determine the suboptimal paths for the second objective function and for $l^a = 7$, the set of Pareto optimal paths is contained in the cone C_2^7 (see Fig. 5.7). Note that the increase of l^a causes an increase in the computation time.

The second refers to the min–max optimal path. We may be sure that the path d^* is certainly the min–max optimal if the inequality (5.13) is satisfied. In this case the values of all function relative increments for the path d^* are smaller than the first function relative increment for the path $d^{(l)}$. Thus none of the next suboptimal paths which satisfy (5.12) can replace the path d^*.

Table 5.1   The list of suboptimal paths to be considered in order to find the min–max optimal path for Example 5.1

l	$d^{(l)}$	$\bar{f}(d^{(l)})$	$\bar{z}(d^{(l)})$
1	$\{u_1, u_5, u_8\}$	$[45, 38]^T$	$[0.0, \quad 0.727]^T$
2	$\{u_1, u_5, u_7, u_8\}$	$[48, 43]^T$	$[0.066, 0.954]^T$
3	$\{u_1, u_5, u_6, u_8\}$	$[56, 35]^T$	$[0.244, 0.590]^T$
4	$\{u_1, u_4, u_7, u_8\}$	$[57, 30]^T$	$[0.266, 0.363]^T$
5	$\{u_1, u_2, u_5, u_8\}$	$[67, 37]^T$	$[0.488, 0.681]^T$

Note that using this method to determine only the path d^* we can terminate the procedure if (5.13) is satisfied.

For Example 5.1, Table 5.1 presents the list of suboptimal paths which have to be considered to satisfy (5.13). The path d^* is the fourth on the list.

It should be mentioned here that the min–max approach to the network optimization problem provides the same possibilities as those discussed in Section 4.3.

CHAPTER 6

Optimization examples

In the previous chapters we discussed multicriterion optimization methods illustrating them with simple engineering problems. This chapter is devoted to more complicated problems, and their selection illustrates the use of multicriterion modelling in different fields. The discussion concentrates more on the problem formulation than on the results obtained which could be improved by using other methods, by providing the engineer with a greater choice of optimal solutions, or by adopting special logic to the problem. The reader may find some excellent examples of other engineering multicriterion optimization problems in Haimes and Hall (1974), Koski (1980), Lightner and Director (1981) and Eschenauer (1983).

The first example deals with the optimal distribution of investment means, i.e., the distribution which ensures the maximal growth in the number of all final products produced by a factory. The optimization model is of the linear programming type.

The second example considers the design of machine tool gearboxes in which the minimization model is of non-linear programming type and the optimization problem consists in finding the dimensions of the gearbox which satisfy the geometric and strength constraints and for which the four objective functions are close to their minima.

The third example describes the multicriterion approach to the so-called black-box system for which the first step is to find a formal description of the system which is then used for building the optimization model. The electrodischarge machining process is used as an illustration of this approach.

In the last example the machining process is modelled by means of a network in which each path represents a certain variant of the process. The optimization problem is to find the path for which the cost and time criteria are optimal.

6.1 DISTRIBUTION OF INVESTMENT MEANS

This example considers the problem of a distribution of investment means to be disposed at a given accounting time. The optimal distribution should ensure an appropriate production growth of final products manufactured in a plant. The number of final products is to be maximized and since all of them should be considered, the optimization model is of the multicriterion type.

We denote by

$I = \{1, 2, ..., i, ..., n\}$ – a set of all parts or units produced in a factory.

$J = \{1, 2, ..., j, ..., k\}$ – a set of all final products produced in a factory,

x_j – a number of the jth product.

Different amounts of particular parts or units may be the components of a final product.

We denote by

$[a_{ji}]$ – matrix representing a number of the ith parts which are the components of the jth final product,

$L = \{1, 2, ..., l, ..., m\}$ – a set of departments producing each part or unit,

$[c_{il}]$ – unit labour–consumption matrix, where c_{il} is an execution time required for producing the ith part in the lth department,

$[b_l]$ – a vector describing the annual or monthly standard hours available in the departments.

The constraints which describe the factory production capacity can be written as follows

$$\sum_{i=1}^{n} c_{il} \sum_{j=1}^{k} a_{ji}x_j \leqslant b_l \qquad \text{for } l = 1, 2, ..., m \qquad (6.1)$$

Let us assume that the factory should reach its maximum production rate for final products and that this may be written formally as follows:

$$f_j(\bar{x}) = x_j \longrightarrow \text{max} \qquad \text{for } j = 1, 2, ..., k \qquad (6.2)$$

Hence the optimization problem at the existing production potential is to find maximum values for k linear functions (6.2) under m linear constraints (6.1). To solve the above multicriterion optimization problem we shall use the min–max method.

In the range of decision problems which may occur in the factory two cases should be distinguished.

Case One

The production rates of final products to be achieved are unknown *a priori* and we aim merely at their maximization. In this case we start from searching the ideal vector $f^0 = [f_1^0, f_2^0, ..., f_k^0]^T$ the elements of which represent maximum values of production rates to be obtained under given manufacturing conditions. Note that f_j^0 corresponds to the maximum value of the jth decision variable. Next, using the min−max method we search the solution which ensures minimum values of the relative increments for all objective functions. By using the weighting min−max method we may ascribe a different degree of significance to the specified production rates.

Case Two

The production rates of final products to be obtained are determined by means of market research. Then the values of f_j^0 for $j = 1, 2, ..., k$ are the assumed extremes to be achieved and the remaining part of the procedure is the same as in Case One.

The optimization procedure described above allows finding the optimal solution for one variant of the constraints (6.1). Let us consider the influence of investment means distribution upon the constraints. The realization of the proposed investments, i.e., purchase of the machines and equipments, modernization of machines, technological modifications, etc., causes changes in the matrix $[c_{il}]$ and sometimes also on the vector $[b_l]$. Let us denote by $S = \{1, ..., s, ..., t\}$ a set of all possible variants of the distribution of investment means. Let us assume that we may calculate or estimate changes in labour consumption for each variant. In this way a new matrix $[c_{il}]$ and sometimes a new vector $[b_i]$ are created for the sth variant.

Carrying out the optimization calculations for each variant we obtain maximum production rates for a given variant. By comparing them we may select an optimal variant. The choice may be based either on the intuition of the decision maker or it may be formalized using the min−max approach. If numerous distribution variants of investment means occur, it may be necessary to develop an algorithm for generating different investment variants and employ a formalized approach.

6.1.1 *Numerical Example*

A factory manufactures three final products which include five different parts in quantities described by matrix

$$[a_{ji}] = \begin{bmatrix} 1 & 2 & 5 & 0 & 4 \\ 1 & 0 & 4 & 4 & 0 \\ 1 & 2 & 5 & 2 & 4 \end{bmatrix}$$

Each part is manufactured in six departments and its labour consumption is represented by matrix

$$[c_{il}] = \begin{bmatrix} 10 & 2 & 5 & 20 & 0 & 0 \\ 0 & 1 & 4 & 0 & 0 & 5 \\ 4 & 0 & 0 & 4 & 4 & 0 \\ 0 & 5 & 4 & 0 & 7 & 3 \\ 5 & 2 & 6 & 0 & 7 & 0 \end{bmatrix}$$

Annual available standard hours in each department are described by vector

$$[b_1] = \begin{bmatrix} 2400 \\ 4800 \\ 7200 \\ 3600 \\ 2400 \\ 3600 \end{bmatrix}$$

The constraints can be presented as follows:

For the first department

$$c_{11}a_{11}x_1 + a_{21}x_2 + a_{31}x_3 + c_{21}a_{12}x_1 + a_{22}x_2 + a_{32}x_3 + c_{31}a_{13}x_1$$
$$+ a_{23}x_2 + a_{33}x_3 + c_{41}a_{14}x_1 + a_{24}x_2 + a_{34}x_3 + c_{51}a_{15}x_1 + a_{25}x_2 + a_{35}x_3 \leqslant b_1$$

After substituting the values and simplifying, we obtain

$$50x_1 + 26x_2 + 50x_3 \leqslant 2400$$

Similarly we obtain the constraints for the remaining departments

$$12x_1 + 22x_2 + 22x_3 \leqslant 4800$$
$$37x_1 + 21x_2 + 45x_3 \leqslant 7200$$
$$40x_1 + 40x_2 + 40x_3 \leqslant 3600$$
$$48x_1 + 48x_2 + 62x_3 \leqslant 2400$$
$$10x_1 + 12x_2 + 16x_3 \leqslant 4600$$

Using the simplex algorithm we can find maximal attainable production rates:

$$
\begin{array}{llll}
\max f_1(x) & x_1 = 34 & x_2 = 0 & x_3 = 0 \\
\max f_2(x) & x_1 = 0 & x_2 = 50 & x_3 = 0 \\
\max f_3(x) & x_1 = 0 & x_2 = 0 & x_3 = 38
\end{array}
$$

Owing to the simple forms of the objective functions, their maximal values may be determined only by analysing the constraints. Now if the maximum value of each function is taken as a rate to be obtained, we get the optimal solution using the min–max approach:

$$
x_1^* = 11 \qquad x_2^* = 16 \qquad x_3^* = 10
$$

If labour consumption of producing each part is reduced by a given variant of investment distribution, that is c_{11} from 10 to 8, c_{45} from 7 to 5 and c_{55} from 7 to 5 then the matrix $[c_{il}]$ will be as follows:

$$
[c_{il}] = \begin{bmatrix}
8 & 2 & 5 & 20 & 0 & 0 \\
0 & 1 & 4 & 0 & 0 & 5 \\
4 & 0 & 0 & 4 & 4 & 0 \\
0 & 5 & 4 & 0 & 5 & 3 \\
5 & 2 & 6 & 0 & 5 & 0
\end{bmatrix}
$$

For the same vector $[b_i]$ a new form of constraints is formulated:

$$
\begin{aligned}
48x_1 + 24x_2 + 48x_3 &\leqslant 2400 \\
12x_1 + 22x_2 + 45x_3 &\leqslant 4800 \\
37x_1 + 21x_2 + 45x_3 &\leqslant 7200 \\
40x_1 + 40x_2 + 40x_3 &\leqslant 3600 \\
40x_1 + 40x_2 + 50x_3 &\leqslant 2400 \\
10x_1 + 12x_2 + 16x_3 &\leqslant 3600
\end{aligned}
$$

For these constraints, the maximal production rates of final products are as follows:

$$
\begin{array}{llll}
\max f_1(x) & x_1 = 42 & x_2 = 0 & x_3 = 0 \\
\max f_2(x) & x_1 = 0 & x_2 = 65 & x_3 = 0 \\
\max f_3(x) & x_1 = 0 & x_2 = 0 & x_3 = 41
\end{array}
$$

The optimal solution in the min–max sense is

$$x_1^* = 13 \qquad x_2^* = 20 \qquad x_3^* = 13$$

Other variants of investment distribution may be considered in a similar way till the most appropriate variant for the highest production growth of all final products is selected.

6.2 DESIGN OF MACHINE TOOL GEARBOXES

The problem is formulated as follows:

For the given structure, structural scheme (layout), kinematic parameters on the input and output shafts and power on the input shaft, find a solution, i.e., basic constructional parameters for numbers of teeth, modules and tooth widths, which minimizes the following objective functions simultaneously:

(i) volume of material used for the elements mounted on the shafts,
(ii) maximal peripheral velocity between gears,
(iii) width of the gearbox,
(iv) distance between axes of the input and output shafts,

and satisfies geometric, kinematic and strength constraints.

The second objective function reflects the dynamic behaviour of the gearbox (vibration and noise) when operating at the highest speed. The remaining functions reflect the weight and dimensions of the gear box.

We shall discuss the problem in detail using as an example a lathe gearbox. The kinematic scheme, the structural scheme and the basic data for this gearbox are presented in Fig. 6.1, where

$n_{I,min}$ = minimum rotational velocity of the input shaft = $280\,\mathrm{rev/min}$,
$n_{I,max}$ = maximum rotational velocity of the input shaft = $1400\,\mathrm{rev/min}$,
$n_{III,min}$ = minimum rotational velocity of the output shaft = $56\,\mathrm{rev/min}$,
N = power on the input shaft = $14.5\,\mathrm{kW}$,
η_I, η_{II} = efficiencies of the mechanisms on shafts I and II ($\eta_I = 1.0$,
$\eta_{II} = 0.97$).

The vector of design variables is

$$x = [x_1, x_2, x_3, x_4, x_5, x_6, x_7, x_8, x_9, x_{10}, x_{11}, x_{12}, x_{13}, x_{14}]^T,$$

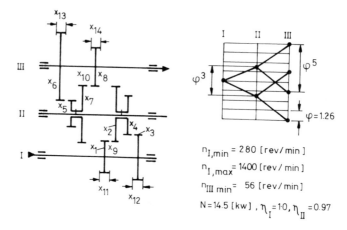

$$n_{I,min} = 280 \; [rev/min]$$
$$n_{I,max} = 1400 \; [rev/min]$$
$$n_{III\;min} = 56 \; [rev/min]$$
$$N = 14.5 \; [kw] \;, \; \eta_I = 1.0, \; \eta_{II} = 0.97$$

Toothed wheel materials – 15 HN carburized and hardened steel.
Type of output shaft – Spindle
Type of modules – Only the commonly used modules.

Fig. 6.1 Kinematic and structural scheme of the gearbox

where

$x_1, x_2, x_3, x_4, x_5, x_6, x_7, x_8$	= number of teeth of the gear wheels,
x_9, x_{10}	= modules of the gear wheels of the double-shaft assemblies,
$x_{11}, x_{12}, x_{13}, x_{14}$	= tooth widths.

Numbers of teeth can assume only integer values, and modules can assume only values from the given set. Knowing the power at the input shaft and the minimum rotational velocities of the input and output shafts, we can estimate the set of modules which will satisfy the strength conditions. For this example we have the following set: $\{2.5, 3.0, 4.0, 5.0, 6.0, 8.0\}$. Modules and tooth widths will be given in mm. We assume that we can use as gear material carburized and hardened steel with the symbol 15 HN, for which permissible bending stress $= 393 \, N/mm^2$, permissible pressures per unit area $= 44.2 \, N/mm^2$ and maximum value of equivalent load factor of teeth $= 2.3$. We also assume the diameters of the shafts: $d_I = 46 \, mm$, $d_{II} = 60 \, mm$ and $d_{III} = 100 \, mm$.

The method for the calculation of strength conditions for the gears is taken from Wrotny (1973), which is adapted from Niemann (1960). For this method and for the above data, the set of constraints with comments, is shown in Table 6.1. The constraints presented are typical for machine tool gearboxes.

Table 6.1 Geometric and strength constraints for the gearbox design problem

Constraints	Comments
$x_1 - 14 \geqslant 0$, $x_2 - 14 \geqslant 0$, $x_3 - 14 \geqslant 0$, $x_4 - 14 \geqslant 0$, $x_5 - 14 \geqslant 0$, $x_6 - 14 \geqslant 0$.	Constraints that ensure the numbers of teeth of all the gear wheels are larger than 14.
$x_1 + x_2 - 35 \geqslant 0$, $100 - x_1 - x_2 \geqslant 0$, $x_3 + x_4 - 35 \geqslant 0$, $100 - x_3 - x_4 \geqslant 0$, $x_5 + x_6 - 35 \geqslant 0$, $100 - x_5 - x_6 \geqslant 0$, $x_7 + x_8 - 35 \geqslant 0$, $100 - x_7 - x_8 \geqslant 0$	Constraints that ensure the sum of teeth of the gear wheels are contained within the range $[35, 100]$
$x_1 + x_2 - x_3 - x_4 = 0$, $x_5 + x_6 - x_7 - x_8 = 0$	Constraints that ensure the same sums of teeth on the transmission gears being components of the double-shaft assembly
$2 - \dfrac{x_1}{x_2} \geqslant 0$, $2 - \dfrac{x_3}{x_4} \geqslant 0$, $2 - \dfrac{x_5}{x_6} \geqslant 0$, $2 - \dfrac{x_7}{x_8} \geqslant 0$, $\dfrac{x_1}{x_2} - 0.25 \geqslant 0$, $\dfrac{x_3}{x_4} - 0.25 \geqslant 0$, $\dfrac{x_5}{x_6} - 0.25 \geqslant 0$, $\dfrac{x_7}{x_6} - 0.25 \geqslant 0$	Constraints that ensure the values of the transmission errors are contained within the range $[0.25, 2]$
$51.12 - \dfrac{280 \cdot x_1 \cdot x_5}{x_2 \cdot x_6} \geqslant 0$, $\dfrac{280 \cdot x_1 \cdot x_5}{x_2 \cdot x_6} - 54.88 \geqslant 0$, $183.6 - \dfrac{280 \cdot x_1 \cdot x_7}{x_2 \cdot x_8} \geqslant 0$, $\dfrac{280 \cdot x_1 \cdot x_7}{x_2 \cdot x_8} - 174.4 \geqslant 0$, $114.24 - \dfrac{280 \cdot x_3 \cdot x_5}{x_4 \cdot x_6} \geqslant 0$, $\dfrac{280 \cdot x_3 \cdot x_5}{x_4 \cdot x_6} - 109.76 \geqslant 0$, $362.1 - \dfrac{280 \cdot x_3 \cdot x_7}{x_4 \cdot x_8} \geqslant 0$, $\dfrac{280 \cdot x_3 \cdot x_7}{x_4 \cdot x_8} - 347,9 \geqslant 0$,	Constraints that ensure the values of the transmission errors are contained within the range $[-2\%, 2\%]$
$x_9 \cdot (x_1 - 2.5) - 46 \geqslant 0$, $x_9 \cdot (x_3 - 2.5) - 46 \geqslant 0$, $x_9 \cdot (x_2 - 2.5) - 60 \geqslant 0$, $x_9 \cdot (x_4 - 2.5) - 60 \geqslant 0$, $x_{10} \cdot (x_5 - 2.5) - 60 \geqslant 0$, $x_{10} \cdot (x_7 - 2.5) - 60 \geqslant 0$, $x_{10} \cdot (x_6 - 2.5) - 100 \geqslant 0$, $x_{10} \cdot (x_8 - 2.5) - 100 \geqslant 0$,	Constraints that ensure the manufacture of the gear wheels on the shaft is possible. Values for the diameter of the shafts are given.
$x_9 \cdot (x_1 + x_2) - x_{10} \cdot x_5 - 96 \geqslant 0$ $x_9 \cdot (x_1 + x_2) - x_{10} \cdot x_7 - 96 \geqslant 0$ $x_{10} \cdot (x_5 + x_6) - x_9 \cdot x_2 - 204 \geqslant 0$ $x_{10} \cdot (x_5 + x_6) - x_9 \cdot x_4 - 204 \geqslant 0$	Constraints that ensure the possibility of placing the gear wheels between the shafts
$\dfrac{x_{11}}{x_9} - 6 \geqslant 0$, $\dfrac{x_{12}}{x_9} - 6 \geqslant 0$, $\dfrac{x_{13}}{x_{10}} - 6 \geqslant 0$, $\dfrac{x_{14}}{x_{10}} - 6 \geqslant 0$, $12 - \dfrac{x_{11}}{x_9} \geqslant 0$, $12 - \dfrac{x_{12}}{x_9} \geqslant 0$, $12 - \dfrac{x_{13}}{x_{10}} \geqslant 0$, $12 - \dfrac{x_{14}}{x_{10}} \geqslant 0$,	Constraints that ensure the tooth width factor values are contained within the range $[6, 12]$

Table 6.1 (continued)

Table 6.1 (continued)

Constraints	Comments

$393 - \dfrac{1\,240\,000 \cdot q_1(x_1, x_9) \cdot q_2(x_1)}{x_1 \cdot x_9^2 \cdot x_{11}} \geq 0$

$393 - \dfrac{1\,240\,000 \cdot q_1(x_1, x_9) \cdot q_2(x_2)}{x_1 \cdot x_9^2 \cdot x_{11}} \geq 0$

$393 - \dfrac{1\,240\,000 \cdot q_1(x_3, x_9) \cdot q_2(x_3)}{x_3 \cdot x_9^2 \cdot x_{12}} \geq 0$

$393 - \dfrac{1\,240\,000 \cdot q_1(x_3, x_9) \cdot q_2(x_4)}{x_3 \cdot x_9^2 \cdot x_{12}} \geq 0$

$393 - \dfrac{6\,010\,000 \cdot q_1(x_5, x_{10}) \cdot q_2(x_5)}{x_6 \cdot x_{10}^2 \cdot x_{13}} \geq 0$

$393 - \dfrac{6\,010\,000 \cdot q_1(x_5, x_{10}) \cdot q_2(x_6)}{x_6 \cdot x_{10}^2 \cdot x_{13}} \geq 0$

$393 - \dfrac{1\,870\,000 \cdot q_1(x_7, x_{10}) \cdot q_2(x_7)}{x_8 \cdot x_{10}^2 \cdot x_{14}} \geq 0$

$393 - \dfrac{1\,870\,000 \cdot q_1(x_7, x_{10}) \cdot q_2(x_8)}{x_8 \cdot x_{10}^2 \cdot x_{14}} \geq 0$

Constraints that ensure the conditions of bending strength are satisfied, where the dynamic factor $q_1(x_r, x_s)$ depends on the diameter of the gear wheel and the tooth shape factor $q_2(x_r)$ depends on the number of teeth.

$44.2 - \dfrac{2\,850\,000 \cdot q_1(x_1, x_g) \cdot q_3(_1, x_2)}{x_1^2 \cdot x_g^2 \cdot (x_{11} - 0.5 \cdot x_g)} \geq 0$

$44.2 - \dfrac{2\,850\,000 \cdot q_1(x_1, x_g),) \cdot q_3(x_1, x_2)}{x_1 \cdot x_2 \cdot x_g^2 (x_{11} - 0.5 \cdot x_g)} \geq 0$

$44.2 - \dfrac{2\,850\,000 \cdot q_1(x_3, x_g) \cdot q_3(x_3, x_4)}{x_3^2 \cdot x_g^2 \cdot (x_{12} - 0.5 \cdot x_g)} \geq 0$

$44.2 - \dfrac{2\,850\,000 \cdot q_1(x_3, x_g) \cdot q_3(x_3 \cdot x_4)}{x_3 \cdot x_4 x_g^2 \cdot (x_{11} - 0.5 \cdot x_g)} \geq 0$

$44.2 - \dfrac{13\,700\,000 \cdot q_1(x_5, x_{10}) \cdot q_3(x_5, x_6)}{x_5 \cdot x_6 \cdot x_{10}^2 \cdot (x_{13} - 0.5 \cdot x_{10})} \geq 0$

$44.2 - \dfrac{13\,700\,000 \cdot q_1(x_5, x_{10}) \cdot q_3(x_5, x_6)}{x_6^2 \cdot x_{10}^2 (x_{13} - 0.5 \cdot x_{10})} \geq 0$

$44.2 - \dfrac{4\,070\,000 \cdot q_1(x_7, x_{10}) \cdot q_3(x_7, x_8)}{x_7 \cdot x_8 \cdot x_{10}^2 \cdot (x_{14} - 0.5 \cdot x_{10})} \geq 0$

$44.2 - \dfrac{4\,070\,000 \cdot q_1(x_7, x_{10}) \cdot q_3(x_7, x_8)}{x_8^2 \cdot x_{10}^2 \cdot (x_{14} - 0.5 \cdot x_{10})} \geq 0$

Constraints that ensure the surface strength conditions are satisfied, where the conversion factor $q_3(x_r, x_s)$ depends on the number of teeth of meshing gear wheels. *Note*: The calculation method for strength conditions of the gear wheels is taken from Wrotny (1973). For this method evaluation of the factors $q_1(x_r, x_s), q_2(x_r), q_3(x_r, x_s)$ cannot be expressed as numerical functions.

Now, we express our objectives:

(1) The volume of material used for the gears can be conveniently approximated by the volume of a cylinder with diameter equal to the addendum circle diameter and height equal to the tooth width. Hence,

the first objective function is

$$f_1(\bar{x}) = 7.86 \times 10^{-7}(x_9^2 x_{11}((x_1 + 2)^2 + (x_2 + 2)^2) + x_9^2 x_{12}((x_3 + 2)^2$$
$$+ (x_4 + 2)^2) + x_{10}^2 x_{13}((x_5 + 2)^2 + (x_6 + 2)^2) + x_{10}^2 x_{14}((x_7 + 2)^2$$
$$+ (x_8 + 2)^2)) \, \text{dcm}^3.$$

(2) The maximal peripheral velocity will occur between gears with numbers of teeth x_7 and x_8 when operating at the highest rotational velocity of the output shaft. Hence, the second objective function is

$$f_2(\bar{x}) = 0.0732 \frac{x_3}{x_4} x_7 x_{10} \, \text{m/sec}.$$

(3) For the simplified model of the gearbox the objective function which represents the width of the gearbox can be expressed as

$$f_3(\bar{x}) = 2(x_{11} + x_{12}a + x_{13} + x_{14}) \, \text{mm}.$$

(4) The distance between the axes of the input and output shafts is

$$f_4(\bar{x}) = 0.5(x_9(x_1 + x_2) + x_{10}(x_5 + x_6)) \, \text{mm}.$$

Now, the problem is to find a vector $\bar{x}^*$ which satisfies the constraints from Table 6.1 and which minimizes simultaneously the objective functions $f_1(\bar{x})$, $f_2(\bar{x})$, $f_3(\bar{x})$, $f_4(\bar{x})$. The method of solution of this problem has been presented by Osyczka (1978) and can briefly be described as follows: first we seek the separately attainable minima of the objective function using the single criterion method based on network modelling (see Osyczka (1976)). Next, using the trade-off method we provide the engineer with Pareto optimal solutions, or employing this method to generate automatically the set of Pareto optimal solutions, we seek the min–max optimal solution.

For the gearbox considered the separately attainable minima give the solutions:

For the first objective function
$\bar{x}^{0(1)} = [29, 48, 42, 35, 20, 60, 41, 39, 4.0, 5.0, 36.0, 28.0, 55.0, 30.0]^{\text{T}}$,
$f_1^0 = 9.489 \, \text{dcm}^3$.
For the second objective function
$\bar{x}^{0(2)} = [26, 64, 40, 50, 25, 50, 46, 29, 4.0, 5.0, 36.0, 24.0, 60.0, 55.0]^{\text{T}}$,
$f_2^0 = 13.472 \, \text{m/sec}$.
For the third objective function
$\bar{x}^{0(3)} = [35, 65, 52, 48, 24, 65, 48, 41, 4.0, 5.0, 24.0, 24.0, 45.0, 30.0]^{\text{T}}$,
$f_3^0 = 246.0 \, \text{mm}$.

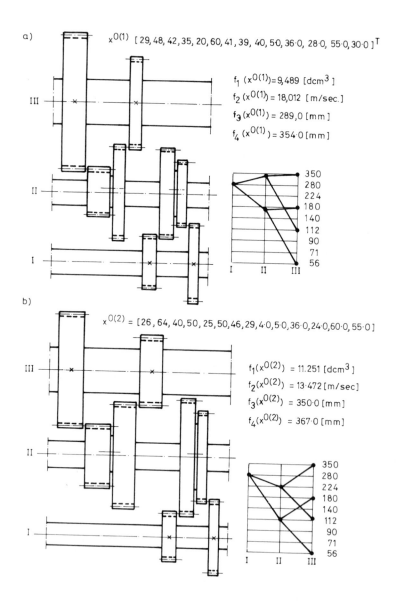

a) $x^{0(1)}$ [29, 48, 42, 35, 20, 60, 41, 39, 40, 5·0, 36·0, 28·0, 55·0, 30·0]^T

$f_1 (x^{0(1)}) = 9,489$ [dcm^3]
$f_2 (x^{0(1)}) = 18,012$ [m/sec.]
$f_3 (x^{0(1)}) = 289,0$ [mm]
$f_4 (x^{0(1)}) = 354·0$ [mm]

b) $x^{0(2)} = $ [26, 64, 40, 50, 25, 50, 46, 29, 4·0, 5·0, 36·0, 24·0, 60·0, 55·0]

$f_1(x^{0(2)}) = 11.251$ [dcm^3]
$f_2(x^{0(2)}) = 13·472$ [m/sec]
$f_3(x^{0(2)}) = 350·0$ [mm]
$f_4(x^{0(2)}) = 367·0$ [mm]

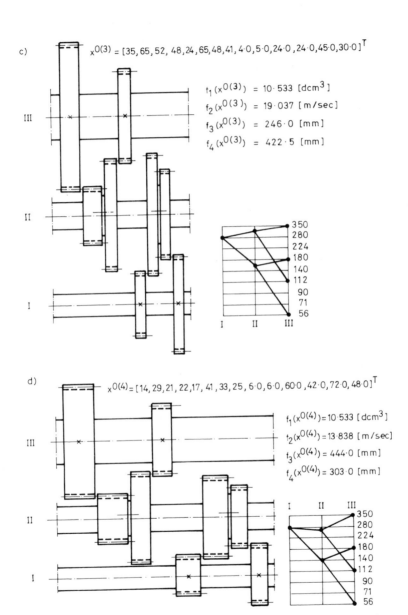

c) $x^{0(3)} = [35, 65, 52, 48, 24, 65, 48, 41, 4 \cdot 0, 5 \cdot 0, 24 \cdot 0, 24 \cdot 0, 45 \cdot 0, 30 \cdot 0]^T$

$f_1(x^{0(3)}) = 10 \cdot 533 \ [dcm^3]$

$f_2(x^{0(3)}) = 19 \cdot 037 \ [m/sec]$

$f_3(x^{0(3)}) = 246 \cdot 0 \ [mm]$

$f_4(x^{0(3)}) = 422 \cdot 5 \ [mm]$

d) $x^{0(4)} = [14, 29, 21, 22, 17, 41, 33, 25, 6 \cdot 0, 6 \cdot 0, 60 \cdot 0, 42 \cdot 0, 72 \cdot 0, 48 \cdot 0]^T$

$f_1(x^{0(4)}) = 10 \cdot 533 \ [dcm^3]$

$f_2(x^{0(4)}) = 13 \cdot 838 \ [m/sec]$

$f_3(x^{0(4)}) = 444 \cdot 0 \ [mm]$

$f_4(x^{0(4)}) = 303 \cdot 0 \ [mm]$

Fig. 6.2 Simplified designs of the gearbox for the separately attainable minima

For the fourth objective function

$$\bar{x}^{0(4)} = [14, 29, 21, 22, 17, 41, 33, 25, 6.0, 6.0, 60.0, 42.0, 72.0, 48.0]^T,$$
$$f_4^0 = 303.0 \text{ mm}.$$

Simplified designs of the gearboxes with speed diagrams for the above solutions are shown in Fig. 6.2.

The min–max optimal solution is:

Vector of the variables

$$\bar{x}^* = [25, 55, 38, 42, 22, 50, 42, 30, 4.0, 6.0, 40.0, 28.0, 48.0, 36.0]^T$$

Vector of the objective functions

$$\bar{f}(\bar{x}^*) = [10.716, 16.694, 304.0, 376.0]^T$$

Vector of the function relative increments

$$\bar{z}(\bar{x}^*) = [0.1293, 0.2391, 0.2357, 0.2409]^T$$

For this solution a simplified design of the gearbox with a speed diagram is shown in Fig. 6.3. All the simplified designs in Fig. 6.2 and Fig. 6.3 are drawn to the same scale. The solution presented in Fig. 6.3. seems to be fully acceptable to the designer.

It should be mentioned here that the min–max approach to the process of designing gearboxes creates the possibility of its automation. This means that we can obtain automatically the solution presented in Fig. 6.3 for the data given in Fig. 6.1. In the computer realization all the constraints which

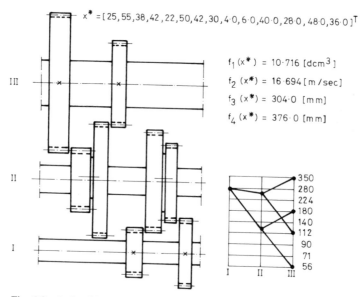

$$x^* = [25, 55, 38, 42, 22, 50, 42, 30, 4.0, 6.0, 40.0, 28.0, 48.0, 36.0]^T$$

$$f_1(x^*) = 10.716 \text{ [dcm}^3\text{]}$$
$$f_2(x^*) = 16.694 \text{ [m/sec]}$$
$$f_3(x^*) = 304.0 \text{ [mm]}$$
$$f_4(x^*) = 376.0 \text{ [mm]}$$

Fig. 6.3 A simplified design of the gearbox for the min–max optimum

are typical are assumed automatically. Tables which contain typical materials used for gears, a range of permitted modules and other data needed for the strength condition calculations are stored in the program. The program also picks out from the range of permitted modules, a set which can satisfy the strength conditions. Of course, there are many other problems connected with the automation of gearbox design which are beyond the scope of this book. We have mentioned this problem only to emphasize that the min–max approach can be an efficient tool in this field.

6.3 ELECTRIC DISCHARGE MACHINING PROCESS

In many engineering activities it may be difficult to obtain an explicitly or implicitly formalized description of a system which could then be optimized. Recently much of work has been devoted to the methods for finding a statistically-experimental model of such systems (see, for example, Volk (1969) and Ogawa (1974)). These methods are based on the theory of experiments and their aim is to determine an investigation programme which is a compromise between a required number of investigations in real life conditions and their informativeness. The data thus obtained are analysed by means of regression methods and the mathematical model of the system is then obtained. This model described the functions dependence between input quantities x_i for $i = 1, 2, ..., n$ and output quantities y_j for $j = 1, 2, ..., k$. Forms of the approximation function can be different. The most frequent forms are:

(a) Linear function

$$y_j = a_0 + \sum_{i=1}^{n} a_i x_i$$

(b) Square function with interactions

$$y_j = a_0 + \sum_{i=1}^{n} a_i x_i + \sum_{\substack{i=1 \\ l>i}}^{n} a_{il} x_i x_l + \sum_{i=l}^{n} a_{ii} x_i^2$$

(c) Power function

$$y_j = a_0 \prod_{i=1}^{n} x_i^{a_i}$$

(d) Logarithmic function with interactions

$$y_j = a_0 \prod_{i=1}^{n} x_i^{a_i + \sum_{l=1}^{n} a_{ij} \ln x_j}$$

Most computer libraries are equipped with standard subroutines which determine the values of a_0, a_i and a_{ij} on the basis of the data obtained from investigations (see, for example, CERN library for CDC computers).

In this way a formal description of the system is obtained. The next step is to build an optimization model which usually has a multicriterion character. In this model the decision variables are those input quantities x_i whose values are to be chosen. Some input quantities may be treated as parameters if their values are fixed in advance. The output quantities y_j form the objectives and constraints and their choice depends on the engineer's requirements. Finally, the results obtained after solving the optimization model provide the answer to the question: which values of input quantities ensure optimal values of output quantities?

As an example of the approach described above we shall consider the Electric Discharge Machining (EDM) process (see Osyczka *et al.* (1982)). In this process the main input quantities x_i are: pulse duration, pulse interval time, amplitude of the discharge current, erosion surface and erosion depth, whereas the output quantities y_j are: metal removal rate, electrode wear, power consumption, surface roughness and dimensional and shape accuracy of the workpiece.

A theoretically recommended approach to the problem of finding a mathematical description of the EDM process would be to carry out investiga-

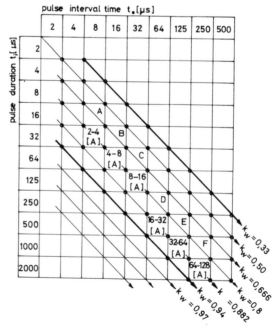

Fig. 6.4 The investigation blocks of the EDM process

tions in the whole region of the space of variables x_i. The space of variables can be, however, restricted to the region which is physically sensible. This region is contained between the lines which are determined by $k_w = 0.50$ and $k_w = 0.94$ (see Fig. 6.4), where

$$k_w = \frac{\text{pulse duration}}{\text{pulse duration} + \text{pulse interval time}}$$

The region is divided into six blocks which are designated by A, B, ..., F in Fig. 6.4. To each block a definite range of variation of the discharge current I in A is assigned. Carrying out investigations according to the assumed programme of the statistical planning of experiments and analysing the results using the regression method a set of equations for each block can be obtained.

These sets of equations give a mathematical description of the EDM process which is the basis for building an optimization model. In this model the decision variables are those input quantities x_i whose values are set on the machine, i.e., pulse duration, pulse interval time and discharge current. If the erosion surface does not change in the machining process and the depth of erosion is given in advance, these two input quantities are treated as parameters.

The choice of the objective and constraint functions depends on the engineer's requirements. The output quantities usually chosen as objective functions are: metal removal rate (maximization) and tool electrode wear (minimization).

The surface roughness and the dimension and shape accuracy of the workpiece can be taken as the third and fourth objective functions in case of the accurate machining. For rough machining these output quantities may be treated as constraints or they may be omitted. Similarly the power consumed by the machine may be either a new objective function or a constraint, or it may be disregarded.

Usually the machine has a final set of possible settings and thus the optimization problem is of the discrete type. The number of decision variables and the small number of their discrete values make it possible to use the systematic search method to generate all points in the space of variables. From all feasible points, i.e., from all those which satisfy constraints, we may select the set of Pareto optimal solutions and/or the min–max optimal solution using the algorithms described in Section 4.1.

We may also work out a computer program which for the given workpiece, i.e., for the given erosion surface and depth, and if required for the assumed surface roughness, automatically selects optimal settings of the machine for the assumed criteria. This program can be applied to a microprocessor controlling the operation of the machine.

Let us now consider a particular instance of the EDM process in detail. We assume that a cylindrical copper electrode is used as the tool and tool steel as the machined material. The results of initial investigation allow us to assume the average working voltage and the dielectric pressure as parameters fixed at values $35^{\pm 3}$ V and 60.8 hPa respectively.

The following input qualities are assumed for the main investigations:

(a) Pulse duration t_i in s,
(b) Pulse interval time t_0 in s,
(c) Discharge current I in A,
(d) Erosion diameter ϕ in mm,
(e) Erosion depth g in mm.

The following output quantities are assumed:

(a) Metal removal rate Q_v in mm^3/min,
(b) Electrode wear δ in per cent,
(c) Surface roughness R_a in m,
(d) Power consumed by the machine N in W.

We shall describe now the results obtained for the block F, see Fig. 6.4. For this block the input quantities range as follows:

$$500 \leqslant t_i \leqslant 2000 \qquad 125 \leqslant t_0 \leqslant 250$$
$$64 \leqslant I \leqslant 128 \qquad 50 \leqslant \phi \leqslant 70$$
$$5 \leqslant g \leqslant 10$$

The dependences between input and output quantities obtained after a series of experiments and their statistical analysis are given below.

$$Q_v = e^{11.744} I^{0.206 + 0.032 \ln t_i + 0.022 \ln t_0 + 0.205 \ln \phi + 0.026 \ln g}$$
$$t_i^{-1.555 + 0.047 \ln t_0 + 2.76 \ln \phi + 0.051 \ln g} t_0^{-0.107 - 0.174 \ln \phi + 0.155 \ln g}$$
$$\phi^{-1.067 - 0.124 \ln g} g^{-0.742}$$

$$\delta = e^{-81.509} I^{5.634 - 0.349 \ln t_i - 0.335 \ln t_0 + 0.119 \ln \phi + 0.174 \ln g}$$
$$t_i^{3.726 - 0.551 \ln t_0 - 0.344 \ln \phi + 0.253 \ln g} t_0^{13.609 - 2.045 \ln \phi + 0.207 \ln g}$$
$$\phi^{12.219 - 0.171 \ln g} g^{-3.102}$$

$$R_a = e^{20.971} I^{-1.687 + 0.224 \ln t_i - 0.027 \ln t_0 + 0.135 \ln \phi - 0.001 \ln g}$$
$$t_i^{-1.199 - 0.085 \ln t_0 - 0.027 \ln \phi - 0.017 \ln g} t_0^{-1.551 + 0.170 \ln \phi + 0.208 \ln g}$$
$$\phi^{-1.967 - 0.345 \ln g} g^{-2.387}$$

$$N = e^{-0.663} I^{1.341 - 0.066 \ln t_i - 0.119 \ln t_0 + 0.140 \ln \phi + 0.053 \ln g}$$
$$t_i^{0.230 + 0.071 \ln t_0 - 0.048 \ln \phi + 0.016 \ln g} t_0^{0.845 - 0.197 \ln \phi - 0.058 \ln g}$$
$$\phi^{0.557 - 0.003 \ln g} g^{0.005}$$

Table 6.2 Optimization results for the workpiece: $\phi = 68$ mm and $g = 6$ mm

	Decision variables			Objective function		
	I (A)	t_i (μs)	t_0 (μs)	Q_v (mm^3/min)	δ (%)	N (W)
Maximum of material removal rate, $Q_v \to$ max	128	2000	125	947.1	0.99	6999.4
Minimum of tool wear, $\delta \to$ min	64	2000	250	420.4	0.19	3508.1
Minimum of power consumption, $N \to$ min	64	500	250	367.6	2.18	2768.2
Optimum in the min–max sense	64	2000	125	455.9	0.22	3623.0

Let us assume the following models of multicriterion optimization.

(i) Decision variables, quantities set on the machine
   pulse duration
   pulse interval time,
   discharge current.

(ii) Objective functions
   metal removal rate,
   electrode wear,
   power consumed by the machine.

Consider the problem of selecting the optimal machining conditions for the workpiece whose diameter $\phi = 68$ mm and depth $g = 6$ mm. For this workpiece and for the above optimization model the results of calculations are shown in Table 6.2. The min–max optimal solution presented in Table 6.2 provides the settings of the machine for which all the three functions are treated as equally important.

6.4 MACHINING PROCESS

The graph modelling of a machining process has been suggested by Szadkowski (1971). In this model the graph $G = \langle u, R \rangle$ represents the set of possible variants of the process. An example of such a graph for the machining process of a shaft is shown in Fig. 6.5. In this graph each arc represents a corresponding machining operation described in Table 6.3, whereas each path represents the feasible sequence of machining operations.

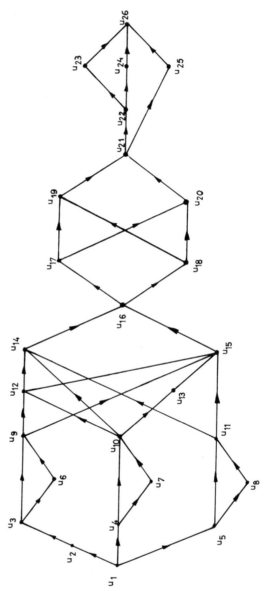

Fig. 6.5 A graph model of the machining process

Table 6.3 Description of machining operations

r	s	Operation	Machine tool
1	2	Manufacture of rod	
1	4	Manufacture of forging	Drop hammer
1	5	Manufacture of forging	Rotary swaging
2	3	Cutting of rod	Circular saw
3	6	Planing	Automatic milling machine
3	9	Planing and centring	Semi-automatic planing and centring machine
6	9	Centring	Automatic centring machine
4	7	like 3–6	
4	10	like 3–9	
7	10	like 6–9	
5	8	like 3–6	
5	11	like 3–9	
8	11	like 6–9	
9	12	Roughing	Multiple-tool semi-automatic lathe
9	15	Roughing and Forming	Semi-automatic tracer-controlled lathe with program control
10	12	Roughing	Multiple-tool semi-automatic lathe
10	14	Roughing and forming	Multiple-spindle chucking machine
10	13	Roughing	Multiple-tool semi-automatic lathe
11	14	Roughing and forming	Multiple-spindle chucking machine
11	15	Roughing and forming	Tracer-controlled lathe
12	14	Forming	Multiple-tool semi-automatic lathe
12	15	Forming	Tracer-controlled lathe
13	14	Forming	Multiple-tool semi-automatic lathe
13	15	Forming	Tracer-controlled lathe
14	16	Rough grinding	Cylindrical grinding machine. Plunge cut-grinding
15	16	Rough grinding	Cylindrical grinding machine. Traverse grinding
16	17	Machining of splines	Multiple-spindle milling machine for splines
16	18	Machining of splines	Broaching machine for splines
17	19	Threading	Milling machine for short thread
17	20	Threading	Lathe with threading head
18	19	like 17–19	
18	20	like 17–20	
19	21	Heat treatment	
20	21	like 19–20	
21	22	Improvement of centre holes	Vertical grinding machine
21	25	Semi-finishing grinding	Centreless grinding machine
22	23	Semi-finishing grinding	Cylindrical grinding machine. Plunge-cut grinding
22	24	Semi-finishing grinding	Cylindrical grinding machine. Traverse grinding
23	26	Finishing grinding	Cylindrical grinding machine
24	26	Finishing grinding	Cylindrical grinding machine
25	26	Finishing grinding	Centreless grinding machine

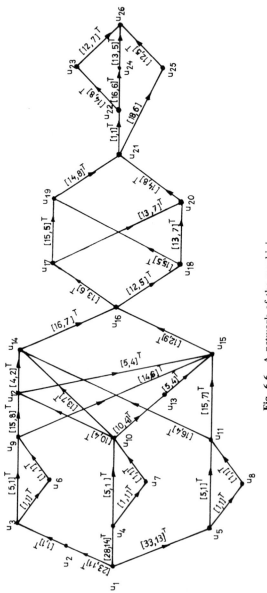

Fig. 6.6 A network of the machining process

To each operation we may ascribe some values which represent the performance criteria of the process. Assuming that these criteria are:

(a) The cost of the operation.
(b) The time of the operation.

we may create the network as presented in Fig. 6.6. In this figure cost and time criteria are reflected only proportionally since the numbers at arcs should be given as integers (see Chapter 5). Note that we do not change the optimization model if we multiply all the values which are associated with the arcs by a number which would yield the integers.

Now the problem is to find the path d^* which represents the optimal sequence of machining operations considering the above criteria.

In this model the path which gives the minimum of the first objective function is

$$d^{0(1)} = \{u_1, u_2, u_3, u_6, u_9, u_{15}, u_{16}, u_{18}, u_{20}, u_{21}, u_{22}, u_{23}, u_{26}\}$$

for which $f_1(d^{0(1)}) = 118$ and $f_2(d^{0(1)}) = 65$.

The path which gives the minimum of the second objective function is

$$d^{0(2)} = \{u_1, u_5, u_{11}, u_{14}, u_{16}, u_{18}, u_{19}, u_{21}, u_{25}, u_{26}\}$$

for which $f_1(d^{0(2)}) = 141$ and $f_2(d^{0(2)}) = 54$.

Using the method of solution discussed in Chapter 5 we provide the engineer with Pareto optimal paths as presented in Table 6.4. These paths have been obtained after considering 150 suboptimal paths for the first objective function. The min–max optimal path is the path for $j = 5$.

Table 6.4 Pareto optimal paths for machining process optimization

j	d_j^p	$f(d_j^p)$	$\bar{z}(d_j^p)$
1	$\{u_1, u_2, u_3, u_6, u_9, u_{15}, u_{16}, u_{18}, u_{20}, u_{21}, u_{22}, u_{23}, u_{26}\}$	$[118, 65]^T$	$[0.0000, 0.2037]^T$
2	$\{u_1, u_2, u_3, u_6, u_9, u_{15}, u_{16}, u_{18}, u_{19}, u_{21}, u_{22}, u_{23}, u_{26}\}$	$[120, 63]^T$	$[0.0169, 0.1666]^T$
3	$\{u_1, u_2, u_3, u_6, u_9, u_{15}, u_{16}, u_{18}, u_{20}, u_{21}, u_{25}, u_{26}\}$	$[121, 60]^T$	$[0.0254, 0.1111]^T$
4	$\{u_1, u_2, u_3, u_6, u_9, u_{15}, u_{16}, u_{18}, u_{19}, u_{21}, u_{25}, u_{26}\}$	$[123, 58]^T$	$[0.0423, 0.0740]^T$
5	$\{u_1, u_2, u_3, u_9, u_{15}, u_{16}, u_{18}, u_{19}, u_{21}, u_{25}, u_{26}\}$	$[126, 57]^T$	$[0.0677, 0.0555]^T$

References

Artobolevski, I. I., Grinkewitch, W. M., Sobol, I. M., Genkin, M. D., Sergeyev, W. I., Statrukov, R. B. (1975), The Method of LP-search for the Optimization of Multiparametric and Multicriterial Problems in Engineering Design, *Computer Aided Design*, 7, 3, 160–162.

Bell, D. E., Keeney, R. L. and Raiffa, H. (eds.) (1977), *Conflicting Objectives in Decisions*, Wiley, New York.

Ben-Tal, A. (1980), Characterization of Pareto and Lexicographic Optimal Solutions. In *Multiple Criteria Decision Making Theory and Application*. G. Fandel and T. Gal, eds., Springer-Verlag, New York.

Boychuk, L. M. and Ovchinnikov, V. O. (1973), Principal Methods of Solution Multicriterial Optimization Problems (survey), *Soviet Automatic Control*, 6, 1–4.

Brayton, R. K. and Spence, R. (1980), *Sensitivity and Optimization*, Elsevier Scientific Publishing Company, Amsterdam.

Carmichael, D. G. (1980), Computation of Pareto Optima in Structural Design, *Int. J. Numer. Methods Eng.*, 15, 6, 925–929.

Chankong, V. and Haimes, Y. Y. (1977), The Interactive Surrogate Worth Trade-off (ISWT) Method for Multiobjective Decision Making. In S. Zionts (ed.) *Multiple Criterion Problem Solving*, Springer-Verlag, New York.

Charnes, A. and Cooper, W. W. (1961), *Management Models and Industrial Applications of Linear Programming*, Vol. I, Wiley, New York. Chapter 6, Appendix B, Basic Existence Theorems and Goal Programming.

Choo, E. U. and Atkins, D. R. (1980), An Interactive Algorithm for Multicriteria Programming, *Comput. and Ops. Res.*, 7, 81–87.

Christofides, N. (1975), *Graph Theory – an Algorithmic Approach*, Academic Press, London and New York.

Cochrane, J. L. and Zeleny, M. (eds.) (1973), *Multiple Criteria Decision Making*, University of South Carolina Press, Columbia, South Carolina.

Cohon, J. L. (1978), *Multiobjective Programming and Planning*, Academic Press, New York.

Colson, G. and Zeleny, M. (1980) Multicriterion Concept of Risk Under Incomplete Information, *Comput. and Ops. Res.*, 7, 125–143.

DaCunha, N. O. and Polak, E. (1967), Constrained Minimization under Vector Valued Criteria in Linear Topological Spaces. In *Mathematical Theory of Control*, Balakrishnan, A. V. and Neustadt, L. W. (eds.), Academic Press, New York.

Dauer, J. P. and Krueger, R. J. (1977), An Interactive Approach to Goal Programming, *Operational Research Quarterly*, 28, 3, 671–681.

Dixon, L. C. W. (1972), *Nonlinear Optimization*, The English Universities Press Limited, London.

Dyer, I. S. (1972), Interactive Goal Programming, Mgmt. Sci., **19**(1), 62–70.

Eschenauer, H. A. (1983), Vector–Optimization in Structural Design and its Application on Antenna Styructures. In *Optimization Methods in Structural Design*, Eschenauer, H. A. and N. Olhoff (eds.), Bibliographisches Institut AG, Zurich.

Fandel, G. and Gal, T. (eds.) (1980), *Multiple Criteria Decision Making Theory and Applications*, Springer-Verlag, New York.

Fishburn, P. C. (1970), *Utility Theory for Decision Making*, Wiley, New York.

Fishburn, P. C. (1974), Lexicographic Order, Utilities and Decision Rules: A Survey, *Mgmt. Sci.*, **20**/11, 1442–1471.

Fletcher, R. and Powell, M. J. D. (1963), A Rapidly Convergent Descent Method for Minimization, *Computer J.*, **6**, 163–171.

Fox, R. L. (1971), *Optimization Methods for Engineering Design*, Addison-Wesley, Reading, Massachusetts.

Gall, D. A. (1966), A Practical Multifactor Optimization Criterion. In *Recent Advances in Optimization Techniques*, Lavi, A. and Vogl, T. P. (eds.), Wiley, New York.

Geoffrion, A, M., Dyer, J. S. and Feinberg, A. (1972), An Interactive Approach for Multicriterion Optimization with an Application to the Operation of an Academic Department, *Management Science*, **19**, 357–368.

Gerlach, Z. (1980), Optymalizacja podstawowych parametrów konstrukcynych i ruchowych przemsłowych skraplaczy ziębniczych chłodzonych wodą. Optimization of Basic Parameters of Water-cooled Condensers. PhD Thesis, Technical University of Cracow, Cracow.

Goncalves, A. S. (1974), A Simplicial Method for Nonlinear Programming. In *Mathematical Programming in Theory and Practice*, Hammer, P. L. and Zoutendijk, G. (eds.), North-Holland, Amsterdam.

Haimes, Y. Y., and Hall, W. A., (1974), Multiobjective in Water Resources Systems Analysis: The Surrogate Worth Trade-off Method, *Wat. Resources Res.*, **10**, 615–624.

Haimes, Y. Y., Hall, W. A. and Freedman, H. T. (1975), *Multiobjective Optimization in Water Resources Systems*, Elsevier, Amsterdam.

Haimes, Y. Y. and Chankong, V. (1979), Kuhn-Tucker Multipliers as Trade-Offs in Multiobjective Decision-Making Analysis, *Automatica*, **15**, 59–72.

Hansen, P. (1980), Bicriterion Path Problems. In *Multiple Criteria Decision Making Theory and Application*, Fandel, G. and Gal, T. (eds.), Springer-Verlag, New York, pp. 109–127.

Himmelblau, D. M. (1972), *Applied Nonlinear Programming*, McGraw-Hill Book Company, New York.

Hooke, R. and Jeeves, T. A. (1961), Direct Search Solution of Numerical and Statistical Problems, *J. Assoc. Comp. Mach.*, **8**, 221–230.

Hu, T. C. (1970), *Integer Programming and Network Flows*, Addison-Wesley, Reading, Massachusetts.

Hwang, C. L., Paidy, S. R., Yoon, K. and Masud, A. S. M. (1980), Mathematical Programming with Multiple Objectives: A Tutorial, *Comp. and Ops. Res.*, **7**, 5–31.

Ignizio, J. P. (1976), *Goal Programming and Extensions*, Lexington Books, Massachusetts.

Ignizio, J. P. (1982), *Linear Programming in Single and Multiple Objective Systems*, Prentice-Hall, Inc., Englewood Cliffs, New Jersey.

Ijiri, Y. (1965), *Management Goals and Accounting for Control*, North-Holland, Amsterdam.

Iwata, K., Murotsu, Y. Iwatsubo, T. and Fuji, S. (1979), A Probabilistic Approach to the Determination of the Optimum Cutting Condition, *Journal of Engineering for Industry, Transaction of American Society of Mechanical Engineer*, **94**, 1099–1107.

Jutler, H. (1967), Liniejnaja modiel z nieskolkimi celevymi funkcjami (Linear Model with Several Objective Functions), *Ekonomika i matematiceckije Metody*, Vol. III, No. 3, pp. 397–406.

Kaczmarek, J. (1976), *Principles of Machining by Cutting, Abrasion and Erosion*, Peter Peregrinus Ltd. Stevenage, WNT, Warsaw.

Keeney, R.L. and Raiffa, H. (1976), *Decision Analysis with Multiple Conflicting Objectives: Preferences and Value Trade-offs*, Wiley, New York.

Korhonen, P. and Soismaa, M. (1981), An Interactive Multiple Criteria Approach to Ranking Alternatives, *J. Opl. Res. Soc.*, **32**, 577–585.

Koski, J. (1979), Optimization with Vector Criterion, *Tampere University of Technology, Publication 6*, Tampere, Finland.

Koski, J. (1980), Truss Optimization with Vector Criterion, Examples, *Tampere University of Technology, Applied Mechanics Report 7*, Tampere, Finland.

Koski, J. (1981), Multicriterion Optimization in Structural Design, *Proceedings of International Symposium on Optimum Structural Design*, University of Arizona, Tucson, Arizona.

Koski, J. and Silvennoinen, R. (1982), Pareto Optima of Isostatic Trusses, *Comp. Methods in Applied Mech. and Eng.*, **31**, 256–279.

Lee, S. M. (1972) *Goal Programming for Decision Analysis*, Auerbach Publishers, Philadelphia, Pennsylvania.

Leitman, G. and Marzollo, A. (1975), *Multicriteria Decision Making*, Springer-Verlag, New York.

Leitman, G. (ed.) (1976), *Multicriteria Decision Making and Differential Games*, Plenum Press, New York.

Leitman, G. (1977), Some Problems of Scalar and Vector-Valued Optimization in Linear Viscoelasticity, *J. Optim. Theory Appl.*, **23**, 1, 93–99.

Lightner, M. R. and Director, S. W. (1981), Multiple Criterion Optimization for the Design of Electronic Circuits, *IEEE Trans. Circuits Syst.*, **28**, 3, 169–179.

Lin, J. G. (1976), Maximal Vector and Multiobjective Optimization, *Journal of Optimization Theory and Application*, **18**, 1, 41–64.

Maas-Teugels, M. (1969), Potentially Critical Paths in Indeterminate Times Scheduling Graph. In *Project Planning by Network Analysis*, Lombaers, H. J. M. (ed.), North-Holland, Amsterdam–London.

Mandl, Ch. (1979), *Applied Network Optimization*, Academic Press, New York.

Marchet, I. C. and Siskos, J. (1979), Aide a la decision en matriere d'environment: Application au choix de trace autoroutier, *Repport LAMSADE, 23–1979*, Universite de Paris Dauphine.

Minami, M. (1981), Weak Pareto Optimality of Multiobjective Problems in a Locally Convex Topological Space, *Journal of Optimization theory and Applications*, **34**, 4, 469–844.

Mistree, F., Heghes, O. F. and Phouc, H. B. (1981), An Optimization Method for the Design of Highly Constrained Complex Systems, *Engineering Optimization*, **5**, 3, 179–197.

Mistree, F. (1983), Design of Damage Tolerant Structural Systems. In *Optimization Methods in Structural Design*, Eschenauer, H. A. and Olhoff, N. (eds.), Bibliographisches Institut AG, Zurich.

Morse, J. N., (ed.) (1981), Organizations: Multiple Agents with Multiple Criteria, *Lecture Notes in Economics and Mathematical Systems No. 190*, Springer-Verlag, New York.

Musselman, K. and Tavalage, J. (1980), A Trade-off Cut Approach to Multiple Objective Optimization, *Operations Research*, **28**, 6, 1424–1435.

Niemann, G. (1960), *Maschinenelemente 1 and 2*, Springer-Verlag, New York.

Ogawa, J. (1974), *Statistical Theory of the Analysis of Experimental Designs*. Marcel Dekker Inc., New York.

Oppenheimer, K. R. (1978), A Proxy Approach to Multi-attribute Decision Making, *Management Science*,24, 675–689.

Osyczka, A. (1975), Optimization of the Steady State Parameters for Machine Tool Gear Trains, *Int. J. Mach. Tool Des. and Res.*, **15**, 31–68.

Osyczka, A. (1976), An Algorithm of Optimization for Special Class of Networks, *Computing*, **16**, 77–97.

Osyczka, A. (1978), An Approach to Multicriterion Optimization Problems for Engineering Design, *Comp. Methods in Applied Mech. and Eng.*, **15**, 309–333.

Osyczka, A. (1979), The Min–max Approach to a Multicriterion Network Optimization Problem, *Proceedings of International Symposium on Mathematical Theory of Networks and Systems*, Delft University of Technology, The Netherlands.

Osyczka, A. (1980), Multicriterion Network Optimization Problem, *Computing*, **25**, 363–368.

Osyczka, A. (1981), An Approach to Multicriterion Optimization for Structural Design, Proceedings of International Symposium on Optimum Structural Design, University of Arizona.

Osyczka, A., Zimny, J., Zając, J. and Bielut, M. (1982), An Approach to Identification and Multicriterion Optimization of EDM Process, *Proceedings of 23rd MTDR Conference*, UMIST, Manchester.

Passy, U. and Levanon, Y. (1980), Manpower Allocation with Multiple Objectives – The Min–Max Approach in Multiple Criteria Decision Making Theory and Application, Fandel, G. and Gal, T. (eds.), Springer-Verlag, New York.

Peschel, M. and Riedel, C. (1976), Polyoptimierung eine Entscheidungshilfe für ingenieurtechnische Kompromisslosungen, VEB Verlag Technik, Berlin.

Payne, J. H. and Polak, E. (1980), An Interactive Rectangle Elimination Method for Biobjective Decision Making, *IEEE Trans. Automatic Control*, **AC–25**, 3, 421–432.

Rao, S. S., and Hati, S. K. (1979), Game Theory Approach in Multicriteria Optimization of Function Generating Mechanisms, *J. Mech. Des. Trans. ASME* **101**, 398–405.

Rosinger, E. E. (1981), Interactive Algorithm for Multiobjective Optimization, *Journal of Optimization Theory and Applications*, **35**, 3, 339–365.

Roy, B. (1971), Problems and Methods with Multiple Objective Functions, *Math Program.*, **1**, 239–266.

Sakawa, M. (1978), Multiobjective Reliability and Redundancy Optimization of a Series–Parallel System, by the Surrogate Worth Trade-off Method, *Microelectronics and Reliability*, **17**, 4, 465–467.

Sakawa, M. (1978), Multiobjective Optimization by the Surrogate Worth Trade-off Method, *IEEE Transactions on Reliability*, **R–27**, 5, 311–314.

Sakawa, M. and Seo, F. (1980), Interactive Multiobjective Decisionmaking for Large-Scale Systems and its Application to Environmental Systems, *IEEE Transaction on Systems, Man and Cybernetics*, **SMC–10**, 12, 796–806.

Sakawa, M. (1980), Multiobjective Optimization for a Standby system by the

Surrogate Worth Trade-off Method, *Journal of the Operational Research Society*, **31**, 2, 153–158.

Sakawa, M. (1981), Interactive Multiobjective Reliability Design of a Standby System by the Sequential Proxy Optimization Techniques (SPOT), *Int. J. Systems Sci.*, **12**, 6, 667–674.

Sakawa, M. (1981), An Interactive Computer Program for Multiobjective Decision Making by the Sequential Proxy Optimization Technique, *Int. J. Man-Machine Studies*, 14, 193–213.

Sakawa, M. and Seo, F. (1982), Interactive Multiobjective Decision Making in Environmental Systems Using Sequential Proxy Optimization Techniques (SPOT), *Automatica*, **12**, 2, 155–165.

Salukvadze, M. E. (1974), On the Existence of Solution in Problems of Optimization under Vector Valued Criteria, *Journal of Optimization Theory and Applications*, **12**, 2, 203–217.

Salukvadze, M. E. (1975), *Zadaci vektornoj optimizacii v teorii upravlenia* (Exercises of Vector Optimization in Control Theory), Mecniereba, Tbilisi.

Salukvadze, M. E. (1979), *Vector-valued Optimization Problems in Control Theory*, Academic Press, New York.

Seinfeld, J. H. and McBride, W. L. (1970), Optimization with Multiple Performance Criteria, *Ind. Eng. Chem. Process Des. Develop.*, **9**, 1, 53.

Siddall, J. N. (1972), Analytical Decision-Making in Engineering Design, Prentice-Hall, Englewood Cliffs, New York.

Solich, R. (1969), Zadanie programowania liniowego z wieloma funkcjami celu (Linear Programming Problem with Several Objective Functions), *Przeglad Statystyczny*, **16**, 24–30.

Stadler, W. (1975), Natural Structural Shapes. In *Multicriteria Decision Making*, Leitmann, G. and Marzollo, A. (eds.), Springer-Verlag, New York.

Stadler, W. (1977), Natural Structural Shapes of Shallow Arches, *J. Appl. Mech. Trans. ASME*, **44**, 2, 291–298.

Stadler, W. (1978), Natural Structural Shapes (The Static Case), *Q. J. Mech. Appl. Math.*, **31**, Pt. 2, 169–217.

Stadler, W. (1979), A Survey of Multicriteria Optimization or the Vector Maximum Problem, Part I: 1776–1960, *Journal of Optimization Theory and Applications*, **29**, 1, 1–52.

Stadler, W. (1981), Stability Implications, and the Equivalence of Stability and Optimality Conditions in the Optimal Design of Uniform Shallow Arches, Proceedings, Symposium on Structural Optimization, 11th Naval Structural Mechanics Symposium, University of Arizona, Tucson, Arizona.

Starr, M. K. and Zeleny, M. (eds.) (1978), *Multiple Criteria Decision Making*, North-Holland, New York.

Steenbrink, P. (1974), *Optimization of Transport Networks*, Wiley, New York.

Steuer, R. E. and Schuler, A. T. (1978), An Interactive Multiple Objective Linear Programming Approach to a Problem in Forest Management, *Op. Res.*, **26**, 254–269.

Szadkowski, J. (1971), An Approach to Machining Process Optimization, *Int. J. Prod. Res.*, **9**, 3, 371–376.

Thiriez, H. and Zionts, S. (1976), Multiple Criteria Decision Making, *Lecture Notes in Economics and Mathematical Systems No. 130*, Springer-Verlag, New York.

Vincent, T. L. and Grantham, W. J. (1981), *Optimality in Parametric Systems*, Wiley, New York.

Vincke, Ph. (1974), *Problemes Multicriteres*, Cahiers du Centre d'Etudes de Recherche Operationnalle, Vol. 16, pp, 425–439.

Volk, W. (1969), *Applied Statistics for Engineers*, McGraw-Hill Inc., New York.

Wallenius, J. (1975), Interactive Multiple Criteria Decision Methods: An Investigation and an Approach, Acta Academiae Oeconomicae Helsingiensis, Seria A: 14, Helsinki School of Economics, Finland.

Wallenius, J. (1975), Comparative Evaluation of Some Interactive Approaches to Multicriterion Optimization, *Mgmt. Sci.*, **21**, 12, 1387–1969.

Walz, F. M. (1967), An Engineering Approach: Hierarchical Optimization Criteria, *IEEE Trans. Automatic Control*, **AC–12**, 179.

White, D. J. (1976), *Fundamentals of Decision Theory*, North-Holland, Amsterdam.

Wierzbicki, A. P. (1975), Penalty Methods in Solving Optimization Problems with Vector Performance Criteria, *Proceedings of VI-th IFAC World Congress*, Cambridge/Boston.

Wierzbicki, A. P. (1978), On the Use of Penalty Functions in Multiobjective Optimization, *Proceedings of the International Symposium on Operations Research*, Mannheim.

Wierzbicki, A. P. (1980), The Use of Reference Objectives in Multiobjective Optimization. In *Multiple Criteria Decision Making Theory and Application*, Fandel, G. and Gal, T. (eds.), Springer-Verlag, New York, 469–486.

Wismer, D. A. and Chattergy, R. (1978), *Introduction to Nonlinear Programming*, North-Holland, New York.

Wrotny, L. T. (1973), *Podstawy budowy obrabiarek*, Basis of Machine Tool Design, WNT, Warsaw.

Yu, P. L. and Leitmann, G. (1974), Compromise Solutions, Domination Structures, and Salukvadze's Solution, *Journal of Optimization Theory and Applications*, **13**, 3, 362–378.

Yu, P. L. and Zeleny, M. (1975), The set of All Nondominated Solutions in Linear Cases and a Multicriteria Simplex Method, *J. Math. Anal. Appl.*, **49**, 2, 430–468.

Zadech, L. A. (1963), Optimality and Non-Scalar Valued Performance Criteria, *IEEE Trans. Automatic Control*, **AC–8**, 59.

Zeleny, M. (1974), A Concept of Compromise Solutions and the Method of the Displaced Ideal, *Computers and Operations Research*, **1**, 4, 479–496.

Zeleny, M. (1974), *Linear Multiobjective Programming*, Springer-Verlag, New York.

Zionts, S. and Wallenius, J, (1976), An Interactive Programming Method for Solving the Multiple Criteria Problem, *Mgmt. Sci.*, **22**, 6, 632–663.

Zionts, S. (ed.) (1978), Multiple Criteria Problem Solving, *Lecture Notes in Economics and Mathematical Systems No. 155*, Springer-Verlag, New York.

Zoutendijk, G. (1974), On Linearly Constrained Nonlinear Programming and Some Extensions. In *Mathematical Programming in Theory and Practice*, Hammer, P. L. and Zoutendijk, G. (eds.), North-Holland, Amsterdam.

A FORTRAN program for random search methods

PURPOSE

To solve the multicriterion optimization problem for non-linear models with inequality constraints and with continuous discrete or mixed valued variables using Methods 1 and 2 described in Section 4.2.

STRUCTURE OF THE PROGRAM

The program consists of the main routine and six subroutines PARETO, MINMAX, ODCHYL, GEN, OGRN and FCEL.

(1) The main routine organizes input–output data, generates the set of feasible solutions and seeks the ideal vector $\bar{f}^0$.

(2) Subroutine PARETO selects the set of Pareto optimal solutions from a set of feasible solutions. The dummy arguments correspond to the symbols from Fig. 4.3 as follows

$$K = k, \ N = n, \ JA = j^a, \ X = \bar{x}^{(l)}, \ F = \bar{f}(\bar{x}^{(l)}), \ XP = \bar{x}^p_j, \ FP = \bar{f}^p_j.$$

(3) Subroutine MINMAX selects the min–max optimal solution from a set of feasible solutions. The dummy arguments correspond to the symbols from Fig. 4.4. as follows

$$K = k, \ N = n, \ L = l. \ LZ = l^*, \ FO = \bar{f}^0, \ X = \bar{x}^{(l)}, \ F = \bar{f}(\bar{x}^{(l)}),$$

$$XOZ = \bar{x}^*, \ FOZ = \bar{f}(\bar{x}^*), \ DFZ = \bar{z}(\bar{x}^*), \ IDZ = v^*_r \times lO^{ICA}.$$

The subroutine compares the function relative increments with an accuracy corresponding to the number of decimal places set by the variable ICA.

(4) Subroutine ODCHYL calculates the function relative increments for MINMAX.
(5) Subroutine GEN generates random values of X(I). Function RANF(n) is a random number generator (see Extended FORTRAN for CDC computers).
(6) Subroutine FCEL and OGRN introduce the optimization problem into the program (the user supplied subroutines).

INTRODUCTION OF THE PROBLEM

The problem is introduced into the program by means of the following subroutines

(1) Objective functions

```
SUBROUTINE  FCEL  (N,KX,FA)
DIMENSION  X(N),FA(K)
FA(1)  =  f₁(x̄)
. . . . . . .
FA(K)  =  fₖ(x̄)
RETURN
END
```

(2) Inequality constraints

```
SUBROUTINE  OGRN  (N,M,X,G)
DIMENSION  X(N),G(M)
G(1)=g₁(x̄)
. . . . . .
G(M)=gₘ(x̄)
RETURN
END
```

The objective and constraint function may be defined by

(1) A simple arithmetic FORTRAN statement such as presented in the program listing.
(2) A more complex analysis which may be in one or more separate subroutines.

INPUT VARIABLES

First card
N = number of decision variables
M = number of inequality constraints
K = number of objective functions

ILA = number of generated points

ICA = accuracy of comparing the function relative increments

$$\text{MET} = \begin{cases} 1 - \text{Method 1} \\ 2 - \text{Method 2} \end{cases}$$

IPR = prints every IPR solution in the feasible region; set IPR equal to zero for no intermediate output

Second card

IX(J) = number of discrete values which the Jth variable can assume; set to one if the Jth variable is continuous valued

Third and the following cards

If IX(J) = 1

XP(J) = estimated lower bound for variable X(J)

XK(J) = estimated upper bound for variable X(J).

If IX(J) > 1

XA(J,I) = discrete values of X(J) variable, where I = 1, 2, ..., IX(J).

OUTPUT VARIABLES

Program prints input data and the results with comments. After the program listing an output for Example 2.2 is shown.

```
          PROGRAM MCM(INPUT,OUTPUT,TAPE1=INPUT,TAPE3=OUTPUT)
C
C
C         PROGRAM MCM
C         MULTICRITERION OPTIMIZATION FOR NONLINEAR PROGRAMMING
C         MONTE CARLO METHODS
C         THIS ROUTINE GENERATES A SET OF FEASIBLE SOLUTIONS
C         AND SEEKS THE SEPARATELY ATTAINABLE MINIMA
C
          DIMENSION FO(10),XO(10,10),F(10),X(10),FOZ(10),XOZ(10),DFZ(10),
         1  DH(10),IDH(10),IB(10),IX(10),XA(10,100),XP(10),XK(10),
         2  IDZ(10),G(10),XPP(10,400),FP(10,400),OD(10,10)
          READ(1,100)N,M,K,ILA,ICA,MET,IPR
          MU=0
          JA=1
          DO 18 I4=1,K
          FP(I4,1)=1.E15
       18 FO(I4)=1.E15
          IDZ(1)=100*10**ICA
          WRITE(3,101)N,M,K,ILA,ICA,MET,IPR
          READ(1,100) (IX(J),J=1,N)
          WRITE(3,306)(IX(J),J=1,N)
          DO 32 J=1,N
          II=IX(J)
          IF(II.EQ.1) GO TO 31
          READ(1,102) (XA(J,I),I=1,II)
          WRITE(3,103) J,(XA(J,I),I=1,II)
```

```
      XP(J)=XA(J,1)
      XK(J)=XA(J,II)
      GO TO 32
   31 READ(1,102) XP(J),XK(J)
      WRITE(3,104)J,XP(J),J,XK(J)
   32 CONTINUE
      WRITE(3,112)
      IF(IPR) 33,33,34
   34 WRITE(3,119)
   33 II=1
   11 CALL GEN(N,X,XA,XP,XK,IX,II)
      IF(M.EQ.0) GO TO 1
      CALL OGRN(N,M,X,G)
      I1=1
    2 IF(G(I1).LT.0) GO TO 10
      I1=I1+1
      IF(I1.LE.M) GO TO 2
    1 CALL FCEL(N,K,X,F)
      I1=1
    4 IF(F(I1).GE.FO(I1)) GO TO 5
      FO(I1)=F(I1)
      DO 6 I3=1,N
    6 XO(I1,I3)=X(I3)
    5 I1=I1+1
      IF(I1.LE.K) GO TO 4
      IF(IPR) 211,211,21
   21 MU=MU+1
      IF(MU/IPR*IPR.NE.MU) GO TO 211
      WRITE(3,107) II,(F(J),J=1,K)
      WRITE(3,114) (X(I5),I5=1,N)
  211 CONTINUE
      IF(MET.EQ.1) GO TO 10
      CALL PARETO(K,N,JA,X,F,XPP,FP)
      IF(JA.EQ.400) GO TO 3
   10 II=II+1
      IF(II.LE.ILA) GO TO 11
      GO TO 8
    3 WRITE(3,113)
    8 DO 7 I=1,K
      WRITE(3,230)I
      DO 77 I1=1,N
   77 X(I1)=XO(I,I1)
      CALL FCEL(N,K,X,F)
      DO 80 IV=1,K
      CALL ODCHYL(IV,K,DH,F,FO)
   80 OD(I,IV)=DH(IV)
      WRITE(3,231)
      DO 250 I5=1,K
  250 WRITE(3,105) I5,F(I5)
      CALL OGRN(N,M,X,G)
      WRITE(3,202)
      DO 25 III=1,M
   25 WRITE(3,201) III,G(III)
      WRITE(3,110)
      DO 7 I1=1,N
    7 WRITE(3,106) I,I1,XO(I,I1)
      WRITE(3,117)
      DO 81 I=1,K
   81 WRITE(3,118) (OD(I,IV),IV=1,K)
      IF(MET.EQ.2) GO TO 190
      II=1
  400 CALL GEN(N,X,XA,XP,XK,IX,II)
      IF(M.EQ.0) GO TO 19
      CALL OGRN(N,M,X,G)
      I1=1
```

```
 12 IF(G(I1).LT.0) GO TO 50
    I1=I1+1
    IF(I1.LE.M) GO TO 12
 19 CALL FCEL(N,K,X,F)
    CALL MINMAX(K,N,II,LZ,FO,X,F,XOZ,FOZ,DFZ,IDZ,ICA)
    IF(IDZ(1).EQ.0) GO TO 40
 50 II=II+1
    IF(II.LE.ILA) GO TO 400
    GO TO 40
190 IF(IPR)49,49,24
 24 WRITE(3,218)
    DO 217 I=1,JA
    WRITE(3,107) I,(FP(J,I),J=1,K)
217 WRITE(3,114)(XPP(J,I),J=1,N)
 49 CONTINUE
    DO 20 II=1,JA
    DO 209 I=1,K
209 F(I)=FP(I,II)
    DO 208 J=1,N
208 X(J)=XPP(J,II)
    CALL MINMAX(K,N,II,LZ,FO,X,F,XOZ,FOZ,DFZ,IDZ,ICA)
    IF(IDZ(1).EQ.0) GO TO 40
 20 CONTINUE
 40 WRITE(3,108) LZ
    WRITE(3,231)
    DO 43 J=1,K
 43 WRITE(3,109) J,FOZ(J)
    WRITE(3,115)
    DO 44 J=1,K
 44 WRITE(3,116)J,DFZ(J)
    WRITE(3,202)
    CALL OGRN(N,M,XOZ,G)
    DO 26 III=1,M
 26 WRITE(3,201) III,G(III)
    WRITE(3,110)
    DO 42 I=1,N
 42 WRITE(3,111) I,XOZ(I)
    STOP
100 FORMAT(8I10)
101 FORMAT(1H1,12X,33H***MULTICRITERION OPTIMIZATION***/
   1    19X,21HRANDOM SEARCH METHODS//17X,24H*********DATA*********/
   2    40H NUMBER OF DECISION VARIABLES...........,I10/
   3    40H NUMBER OF INEQUALITY CONSTRAINTS.......,I10/
   4    40H NUMBER OF OBJECTIVE FUNCTIONS..........,I10/
   5    40H NUMBER OF GENERATED POINTS.............,I10/
   6    40H ACCURACY OF COMPARING OF INCREMENTS....,I10/
   7    40H METHOD.................................,I10/
   8    40H INTERMEDIATE OUTPUT....................,I10/)
102 FORMAT(8F10.4)
103 FORMAT(/4H XA(,I2,6H,I) = ,(4E15.5))
104 FORMAT(4H XP(,I2,3H) =,E15.5,5X,3HXK(,I2,3H) =,E15.5)
105 FORMAT(3H F(,I2,2H)=,E15.5)
106 FORMAT(4H XO(,I2,1H,,I2,4H) = ,E15.5)
107 FORMAT(6H II = ,I5,14H    VECTOR F = ,(4E15.5))
108 FORMAT(//30H  THE MIN-MAX OPTIMAL SOLUTION /5H LZ =,I5)
109 FORMAT(5H FOZ(,I2,3H)= ,E15.5)
110 FORMAT(26H  DECISION VARIABLE VALUES)
111 FORMAT(5H XOZ(,I2,4H) = ,E15.5)
112 FORMAT(/17X,25H*********RESULTS********* )
113 FORMAT(56H NUMBER OF PARETO OPTIMAL SOLUTIONS HAS EXCEEDED ARRAY D
   1    9HIMENSIONS )
114 FORMAT(12H VECTOR X = ,(4E15.5))
115 FORMAT(30H  FUNCTION RELATIVE INCREMENTS )
116 FORMAT(5H DFZ(,I2,3H) =,E15.5)
```

```
117 FORMAT(1X,45(1H-)/
  1   46H PAYOFF TABLE FOR FUNCTION RELATIVE INCREMENTS /1X,45(1H-))
118 FORMAT(1X,8E15.5)
119 FORMAT(/33H SOLUTIONS IN THE FEASIBLE REGION)
201 FORMAT(3H G(,I2,2H)=,E15.5)
202 FORMAT(30H  INEQUALITY CONSTRAINT VALUES)
218 FORMAT(//3X,24HPARETO OPTIMAL SOLUTIONS )
230 FORMAT(11H RESULTS OF,I3,23H  FUNCTION MINIMIZATION )
231 FORMAT(27H  OBJECTIVE FUNCTION VALUES)
306 FORMAT(7H IX(J)=,6I10)
    END
C
C
C     SUBROUTINE PARETO SELECTS THE SET OF PARETO OPTIMAL SOLUTIONS
C
    SUBROUTINE PARETO(K,N,JA,X,F,XP,FP)
    DIMENSION X(10),F(10),XP(10,400),FP(10,400)
    J=1
  4 KA=0
    DO 1 I=1,K
    IF(F(I).LE.FP(I,J)) KA=KA+1
  1 CONTINUE
    IF(KA.EQ.K) GO TO 2
    IF(KA.EQ.0) GO TO 3
    J=J+1
    IF(J.LE.JA) GO TO 4
    JA=JA+1
    DO 5 J1=1,N
  5 XP(J1,JA)=X(J1)
    DO 6 I=1,K
  6 FP(I,JA)=F(I)
    GO TO 3
  2 DO 7 J1=1,N
  7 XP(J1,J)=X(J1)
    DO 8 I=1,K
  8 FP(I,J)=F(I)
  3 RETURN
    END
C
C
C     SUBROUTINE MINMAX SELECTS THE MIN-MAX OPTIMAL SOLUTION
C
    SUBROUTINE MINMAX (K,N,L,LZ,FO,X,F,XOZ,FOZ,DFZ,IDZ,ICA)
    DIMENSION FO(10),F(10),X(10),FOZ(10),XOZ(10),DFZ(10),DH(10),
   1TDH(10),IB(10),ID(10),IDZ(10)
    IND=0
    I2=1
  6 CALL ODCHYL(I2,K,DH,F,FO)
    IDH(I2)=DH(I2)*10**ICA
    IF(IDH(I2).NE.0) GO TO 2
    IND=IND+1
  2 I2=I2+1
    IF(I2.LE.K) GO TO 6
    IF(IND.LT.K) GO TO 3
    DO 4 I4=1,K
    DFZ(I4)=0.0
  4 FOZ(I4)=F(I4)
    DO 5 I3=1,N
  5 XOZ(I3)=X(I3)
    IDZ(1)=0
    LZ=L
    GO TO 99
  3 I3=0
    DO 7 I2=1,K
```

```
    7 IB(I2)=0
   11 I2=1
      IDA=0
    9 IF(IB(I2).NE.0) GO TO 8
      IF(IDA.GT.IDH(I2)) GO TO 8
      IDA=IDH(I2)
      IS=I2
    8 I2=I2+1
      IF(I2.LE.K) GO TO 9
      I3=I3+1
      ID(I3)=IDA
      IB(IS)=1
      IF(I3.LE.K) GO TO 11
      I3=1
   13 IF(ID(I3).LT.IDZ(I3)) GO TO 20
      I3=I3+1
      IF(I3.LE.K) GO TO 13
      GO TO 99
   20 I5=I3-1
      IF(I5.EQ.0) GO TO 12
      IF(ID(I5).EQ.IDZ(I5)) GO TO 20
      GO TO 99
   12 DO 15 I3=1,N
   15 XOZ(I3)=X(I3)
      LZ=L
      DO 16 I4=1,K
      IDZ(I4)=ID(I4)
      DFZ(I4)=DH(I4)
   16 FOZ(I4)=F(I4)
   99 RETURN
      END
C
C
C
C     SUBROUTINE ODCHYL CALCULATES THE FUNCTION RELATIVE INCREMENTS
C
      SUBROUTINE ODCHYL(I,K,DH,F,FO)
      DIMENSION DH(K),F(K),FO(K)
      DH(I)=ABS(F(I)-FO(I))/ABS(FO(I))
      DH1=ABS(F(I)-FO(I))/ABS(F(I))
      IF(DH1.GT.DH(I)) DH(I)=DH1
      RETURN
      END
C
C
C     SUBROUTINE GEN GENERATES RANDOM VALUES OF X(I)
C
      SUBROUTINE GEN(N,X,XA,XP,XK,IX,II)
      DIMENSION X(10),XA(10,100),XP(10),XK(10),IX(10)
      DO 5 J=1,N
      X(J)=XP(J)+RANF(II)*(XK(J)-XP(J))
      IF(IX(J).EQ.1) GO TO 5
      I=1
    2 V=I+1
      IF(X(J).LE.XA(J,V)) GO TO 3
    7 I=I+1
      GO TO 2
    3 IF(X(J).LT.XA(J,I)) GO TO 7
      IF(X(J).LT.XA(J,I)+0.5*(XA(J,V)-XA(J,I))) GO TO 4
      X(J)=XA(J,V)
      GO TO 5
    4 X(J)=XA(J,I)
    5 CONTINUE
      RETURN
      END
```

```
C
C
C     SUBROUTINE OGRN INTRODUCES INEQUALITY CONSTRAINTS INTO PROGRAM
C
      SUBROUTINE OGRN(N,M,X,G)
      DIMENSION X(N),G(M)
      G(1)=12.-X(1)-X(2)
      G(2)=-X(1)*X(1)+10.*X(1)-X(2)*X(2)+16.*X(2)-80.
      RETURN
      END
C
C
C     SUBROUTINE FCEL INTRODUCES OBJECTIVE FUNCTIONS INTO PROGRAM
C
      SUBROUTINE FCEL(N,K,X,FA)
      DIMENSION X(N),FA(K)
      FA(1)=X(1)+X(2)*X(2)
      FA(2)=X(1)*X(1)+X(2)
      RETURN
      END
```

^***

Output example


```
                 ***MULTICRITERION OPTIMIZATION***
                       RANDOM SEARCH METHODS

                 **********DATA**********
NUMBER OF DECISION VARIABLES............         2
NUMBER OF INEQUALITY CONSTRAINTS........         2
NUMBER OF OBJECTIVE FUNCTIONS...........         2
NUMBER OF GENERATED POINTS..............       100
ACCURACY OF COMPARING OF INCREMENTS.....         5
METHOD..................................         2
INTERMEDIATE OUTPUT.....................         0

IX(J)=          6          6

XA( 1,I) =      .20000E+01    .30000E+01    .40000E+01    .50000E+01
       .60000E+01     .70000E+01


XA( 2,I) =      .50000E+01    .60000E+01    .70000E+01    .80000E+01
       .90000E+06     .10000E+02

                 ********RESULTS********
RESULTS OF  1  FUNCTION MINIMIZATION
 OBJECTIVE FUNCTION VALUES
F( 1)=      .30000E+02
F( 2)=      .30000E+02
 INEQUALITY CONSTRAINT VALUES
G( 1)=      .20000E+01
G( 2)=     0.
 DECISION VARIABLE VALUES
XO( 1, 1) =       .50000E+01
XO( 1, 2) =       .50000E+01
RESULTS OF  2  FUNCTION MINIMIZATION
 OBJECTIVE FUNCTION VALUES
F( 1)=      .66000E+02
F( 2)=      .12000E+02
```

```
INEQUALITY CONSTRAINT VALUES
G( 1)=      .20000E+01
G( 2)=    0.
DECISION VARIABLE VALUES
X0( 2, 1) =      .20000E+01
X0( 2, 2) =      .80000E+01
-------------------------------------------------
PAYOFF TABLE FOR FUNCTION RELATIVE INCREMENTS
-------------------------------------------------
    0.              .15000E+01
    .12000E+01    0.


THE MIN-MAX OPTIMAL SOLUTION
LZ =    2
OBJECTIVE FUNCTION VALUES
FOZ( 1)=      .39000E+02
FOZ( 2)=      .15000E+02
FUNCTION RELATIVE INCREMENTS
DFZ( 1) =      .30000E+00
DFZ( 2) =      .25000E+00
INEQUALITY CONSTRAINT VALUES
G( 1)=      .30000E+01
G( 2)=      .10000E+01
DECISION VARIABLE VALUES
XOZ( 1) =      .30000E+01
XOZ( 2) =      .60000E+01
```

A FORTRAN program for the interactive multicriterion optimization system

PURPOSE

To solve a multicriterion optimization problem for non-linear models with inequality and equality constraints and with continuous valued variables.

STRUCTURE OF THE PROGRAM

The structure of the program is shown in Fig. B.1. where:

(1) The main routine controls all the work of the system and its interactive realization.

(2) Subroutines ZMTOL, FEASBL, START and SUMR solve the single criterion optimization problem using the flexible tolerance method.

(3) Subroutines SEKLOS, SZUK, LOSMC, LOS, KARA and GEBNRT solve the single criterion optimization problem using the direct and random search method.

(4) Subroutine TABL calculates and prints the pay-off tables.

(5) Subroutine WRITEX prints the results for the optimal solution.

(6) Subroutines FCEL and FCEL1 calculate a value of minimized quantity for all multicriterion optimization methods employed in the system.

(7) Subroutines MINMAX and ODCHYL calculate the function relative increments for min–max and weighting min–max methods.

(8) Subroutines FCELU, OGRN and OGRR introduce the optimization problem into the program (the user supplied subroutines).

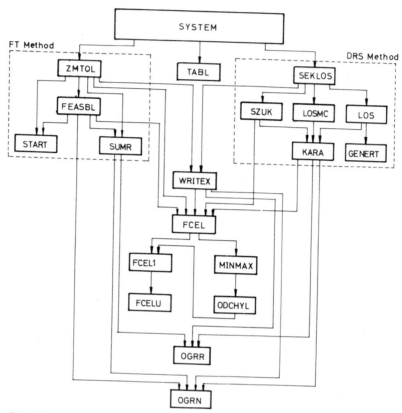

Fig. B.1 The structure of the program for the interactive multicriterion optimization system

The structure of the COMMON blocks is as follows

COMMON/1/ contains common variables used in each single criterion optimization method,

COMMON/2/ contains variables used only in the flexible tolerance method,

COMMON/3/ contains variables used only in the direct and random search method,

COMMON/C/ contains variables used in all multicriterion optimization methods,

COMMON/D/ contains variables for calculating the pay-off tables.

The structure of the program and COMMON blocks enable the user to introduce another single criterion optimization method by adding relevant subroutines instead of or in addition to those which are distinguished by dash-lined blocks in Fig. B.1.

INTRODUCTION OF THE PROBLEM

The problem is introduced into the program by means of the following subroutines

(1) Objective functions

```
SUBROUTINE FCELU(X,FU)
DIMENSION X(20),FU(20)
FU(1)=f₁(x)
. . . . . ..
FU(K)=fₖ(x)
RETURN
END
```

(2) Inequality constraints

```
SUBROUTINE OGRN(X,G)
DIMENSION X(20),G(40)
G(1)=g₁(x)
. . . . ..
G(M)=gₘ(x)
RETURN
END
```

(3) Equality constraints

```
SUBROUTINE OGRR(X,H)
DIMENSION X(20),H(40)
H(1)=h₁(x)
. . . . ..
H(p)=hₚ(x)
RETURN
END
```

The objective and constraint function may be defined in the same way as for the program in Appendix A.

INPUT VARIABLES

The data for input variables are introduced interactively and thus we shall explain only those data which are not simple questions to be answered while running the system.

(1) Common data for the system
NX = number of decision variables,
NIC = number of equality constraints,
NC = number of inequality constraints,
N = number of objective functions,

IDR = $\begin{cases} 0, \text{ no intermediate output,} \\ 1, \text{ intermediate output is printed,} \end{cases}$

XK(I) = estimated upper bounds on x_i, where I = 1, 2, ..., NX,
XP(I) = estimated lower bounds on x_i, where I = 1, 2, ..., NX.

(2) Initial guesses for starting point
XAC(JC) = starting values of x_i, where JC = 1, 2, ..., NX.

(3) Data for the flexible tolerance method
CONVER = an arbitrarily chosen positive small number used to terminate the search which is also known as the convergence criterion,
SIZE = the size of the flexible polyhedron during the initiating phase of the search (see Himmelblau (1972)).
The recommended value for SIZE is as follows
(a) SIZE ≈ 20 per cent of difference between an upper and lower bound on x, if the expected range of variation of the x's along each coordinate is about the same.
(b) If the expected range of variation of the x's along each coordinate is different, make SIZE ≈ to the smallest difference between the respective upper and lower bounds for any x.

(4) Data for the direct and random search method
MAXM = maximum number of moves permitted in the direct search method,
NTEST = number of random points to be generated in the shotgun search in the vicinity of the minimum found by the direct search method,
FPK = fraction of range used as the initial step size,
WK = fraction of the initial step size used as the convergence criterion,
NTMC = number of randomly generated points for finding a new starting point from which the direct search method seeks a better solution,
NRAZ = number of direct search method runs using new starting points,
NH = number of shotgun searches permitted.

The program is equipped with recommended values of input variables for both single criterion optimization methods but the user may introduce other values more suitable for his problem.

OUTPUT VARIABLES

All output variables are printed with comments. The structure of the output is explained in Section 4.4.

```
      PROGRAM SYSTEM(INPUT,OUTPUT)
C
C
C     PROGRAM SYSTEM
C     INTERACTIVE MULTICRITERION OPTIMIZATION SYSTEM FOR NONLINEAR
C     PROGRAMMING
C
      COMMON/1/NX,NC,NIC,X(20),XAC(20),XK(20),XP(20),ID,IDR,R(40)
      COMMON/2/CONVER,SIZE,STEP,ALFA,BETA,GAMA,IN,INF,FDIFER,SEQL,K1,
     1K2,K3,K4,K5,K6,K7,K8,K9,X1(20,20),X2(20,20),SUM(20),SR(20),F(20),
     2ROLD(20),SCALE,FOLD,LFEAS,L5,L6,L7,L8,L9,R1A,R2A,R3A
      COMMON/3/MAXM,NVIOL,FPK,WK,NNDEX,FEX,KRK,NTEST,LW,XV(20),NH,FCE,
     6NTMC,NRAZ,XSTRT(20),A1(20),A2(20),A3(20),A4(20)
      COMMON/C/M,A(20),P(20),N,KK,MIA(20),WSP(40),IS,IPO
      COMMON/D/DW(20,20),DR(20,20),PO(20,20)
      DATA CONVER,SIZE,MAXM,NTEST,FPK,WK,NTMC,NRAZ,NH
     1/0.00001,1.0,200,200,0.1,0.0001,300,2,2/
      PRINT 201
  499 CONTINUE
      PRINT 422
      READ *,NX,NC,NIC,N,IDR
      READ *,(XK(I),I=1,NX)
      READ *,(XP(I),I=1,NX)
      PRINT 203,NX,NC,NIC,N,IDR
      PRINT 310,(XK(I),I=1,NX)
      PRINT 311,(XP(I),I=1,NX)
  397 CONTINUE
      PRINT 423
      READ *,(XAC(JC),JC=1,NX)
      PRINT 312,(XAC(JC),JC=1,NX)
 9999 CONTINUE
      PRINT 401
      PEAD *,IMOJ
      IF(IMOJ.LT.0) GO TO 104
      PRINT 205
      IF(IMOJ-2) 101,102,104
  101 PRINT 301
      PRINT 402
      READ *,CONVER,SIZE
      PRINT 302
      PRINT 303,CONVER,SIZE
      GO TO 103
  102 PRINT 305
      PRINT 402
      READ *,MAXM,NTEST,FPK,WK,NTMC,NRAZ,NH
      PRINT 302
      PRINT 306,MAXM,NTEST,FPK,WK,NTMC,NRAZ,NH
  103 PRINT 205
```

```
104 PRINT 403
    IMOJ=IABS(IMOJ)
    READ *,IEF
    IF(IEF-2)7,4,6
  7 IS=15
    MIA(IS)=1
    ID=1
    M=0
    PRINT 216
    DO 2 J=1,N
    KK=0
    M=M+1
    PRINT 411,M
    IF(IMOJ-2) 1001,1002,1002
1001 CALL ZMTOL
    GO TO 1003
1002 CALL SEKLOS
1003 CONTINUE
    IF(N.EQ.1) GO TO 6
    NTAB=0
    CALL TABL(NTAB)
  2 CONTINUE
    NTAB=1
    CALL TABL(NTAB)
    GO TO 6
  4 PRINT 404
    READ *,(A(JW),JW=1,N)
    PRINT 214,(JW,A(JW),JW=1,N)
  6 IS=0
    M=N
    M=M+1
  1 IS=IS+1
    PRINT 405
    READ *,KIA
    MIA(IS)=KIA
    IF(MIA(IS).EQ.0) GO TO 399
    PRINT 205
    GO TO (151,152,153,154,155),KIA
151 PRINT 551
    GO TO 599
152 PRINT 552
    GO TO 599
153 PRINT 553
    GO TO 599
154 PRINT 554
    GO TO 599
155 PRINT 555
599 PRINT 205
    I=0
 90 I=I+1
    ID=I
    KK=1
    IF(KIA.NE.2) GO TO 5
    PRINT 406
    READ *,IPO
    PRINT 215,I,IPO
  5 IF(KIA.LE.2) GO TO 80
    PRINT 407
    READ *,(WSP(JA),JA=1,N)
    GO TO (80,80,50,60,60),KIA
 50 PRINT 209,I,(JA,WSP(JA),JA=1,N)
    GO TO 80
 60 PRINT 210,I,(JA,WSP(JA),JA=1,N)
 80 IF(IMOJ-2) 1011,1012,1012
```

```
1011 CALL ZMTOL
     GO TO 3
1012 CALL SEKLOS
   3 CONTINUE
     GO TO (131,132,133,133,133),KIA
 132 PRINT 408
     GO TO 134
 133 PRINT 412
 134 READ *,KLP
     IF(KLP.EQ.1) GO TO 90
 131 PRINT 409
     READ *,KZW
     IF(KZW.EQ.1) GO TO 1
     IF(IEF.NE.2) GO TO 399
     PRINT 839
     READ *,KIC
     IF(KIC.EQ.1) GO TO 4
 399 CONTINUE
     PRINT 410
     READ *,IZA
     IF(IZA.EQ.1) GO TO 9999
     PRINT 421
     READ *,IRA
     IF(IRA.EQ.1) GO TO 397
     PRINT 424
     READ *,INA
     IF(INA.EQ.1) GO TO 499
     PRINT 212
     STOP
 201 FORMAT(1H1,72(1H*),/,13X,
    1 46HINTERACTIVE MULTICRITERION OPTIMIZATION SYSTEM,/,21X,
    2 30HTECHNICAL UNIVERSITY OF CRACOW,/,29X,13HCRACOW POLAND,/,
    3 13X,46(1H-))
 203 FORMAT(10X,26HCOMMON DATA FOR THE SYSTEM/10X,
    1 28HNUMBER OF DECISION VARIABLES,17X,I5/10X,
    2 30HNUMBER OF EQUALITY CONSTRAINTS,15X,I5/10X,
    3 32HNUMBER OF INEQUALITY CONSTRAINTS,13X,I5/10X,
    4 29HNUMBER OF OBJECTIVE FUNCTIONS,16X,I5/10X,
    5 19HINTERMEDIATE OUTPUT,26X,I5)
 204 FORMAT(1X,71(1H*))
 205 FORMAT(1X,71(1H-))
 209 FORMAT(9H SOLUTION,I3,/(10X,6HDELTA(,I2,2H)=,F5.2))
 210 FORMAT(9H SOLUTION,I3,/(10X,7HLAMBDA(,I2,2H)=,F5.2))
 212 FORMAT(29H ---END OF JOB,THANK YOU--- )
 214 FORMAT(1X,71(1H-)/10X,37HASSUMED MINIMA OF OBJECTIVE FUNCTIONS/1X,
    1 71(1H-)/(10X,10HMINIMUM OF,I3,23H OBJECTIVE FUNCTION = ,E12.5))
 215 FORMAT(9H SOLUTION,I3,/,5X,11HEXPONENT = ,I5)
 216 FORMAT(1X,71(1H-)/10X,40HCALCULATED MINIMA OF OBJECTIVE FUNCTIONS/
    1 1X,71(1H-))
 310 FORMAT(10X,30HESTIMATED UPPER BOUNDS ON X(I),/(10X,3E16.6))
 311 FORMAT(10X,30HESTIMATED LOWER BOUNDS ON X(I),/(10X,3E16.6))
 301 FORMAT(10X,51HYOU HAVE CHOSEN FT METHOD FOR FUNCTION MINIMIZATION)
 302 FORMAT(10X,19HDATA FOR THE METHOD)
 303 FORMAT(10X,19HDESIRED CONVERGENCE,21X,E12.5/10X
    1 26HSIZE OF INITIAL POLYHEDRON,14X,E12.5)
 305 FORMAT(10X,52HYOU HAVE CHOSEN DRS METHOD FOR FUNCTION MINIMIZATION
    1 )
 306 FORMAT(10X,33HMAXIMUM NUMBER OF MOVES PERMITTED,23X,I6/
    1 10X,39HNUMBER OF TEST POINTS IN SHOTGUN SEARCH,17X,I6/
    2 10X,35HFRACTION OF RANGE USED AS STEP SIZE,15X,E12.5/
    3 10X,48HSTEP SIZE FRACTION USED AS CONVERGENCE CRITERION,2X,E12.5
    4 /10X,47HNUMBER OF RANDOMLY GENERATED POINTS FOR FINDING/
    5 10X,20HA NEW STARTING POINT,36X,I6/
    6 10X,45HNUMBER OF DIRECT SEARCH METHOD RUNS USING NEW/
```

```
      7  10X,15HSTARTING POINTS,41X,I6/
      8  10X,36HNUMBER OF SHOTGUN SEARCHES PERMITTED,20X,I6)
    312 FORMAT(10X,23HSTARTING VALUES OF X(I),/(10X,3E16.6))
    401 FORMAT(39H ---SELECT FUNCTION MINIMIZATION METHOD,2X)
    402 FORMAT(30H ---GIVE DATA FOR THE METHOD    )
    403 FORMAT(37H ---SELECT ONE OF THESE POSSIBILITIES/
      1  43H    1 - MINIMA ARE CALCULATED BY THE SYSTEM/
      2  40H    2 - MINIMA WILL BE GIVEN BY THE USER/
      3  35H    3 - MINIMA ARE NOT CALCULATED   )
    404 FORMAT(27H ---GIVE THE ASSUMED MINIMA,2X)
    405 FORMAT(45H ---SELECT MULTICRITERION OPTIMIZATION METHOD,2X)
    551 FORMAT(29(1H ),14HMIN-MAX METHOD)
    552 FORMAT(25(1H ),23HGLOBAL CRITERION METHOD)
    553 FORMAT(24(1H ),24HWEIGHTING MIN-MAX METHOD)
    554 FORMAT(25(1H ),21HPURE WEIGHTING METHOD)
    555 FORMAT(22(1H ),27HNORMALIZED WEIGHTING METHOD )
    406 FORMAT(27H ---GIVE THE EXPONENT VALUE,2X)
    407 FORMAT(35H ---GIVE THE WEIGHTING COEFFICIENTS,2X)
    408 FORMAT(46H ---DO YOU WANT TO GIVE ANOTHER EXPONENT VALUE,2X)
    412 FORMAT(52H ---DO YOU WANT TO GIVE OTHER WEIGHTING COEFFICIENTS,2X
    409 FORMAT(56H ---DO YOU WANT TO SELECT ANOTHER MULTICRITERION OPTIMIZ
      1  12HATION METHOD,2X)
    410 FORMAT(56H ---DO YOU WANT TO SELECT ANOTHER FUNCTION MINIMIZATION
      1  6HMETHOD,2X)
    411 FORMAT(14H ***RESULTS OF,I3,2X,24HFUNCTION MINIMIZATION***)
    421 FORMAT(53H ---DO YOU WANT TO GIVE NEW STARTING VALUES OF X(I)   )
    422 FORMAT(35H ---GIVE COMMON DATA FOR THE SYSTEM,2X)
    423 FORMAT(32H ---GIVE STARTING VALUES OF X(I),2X)
    424 FORMAT(40H ---DO YOU WANT TO CONTINUE CALCULATIONS,2X)
    839 FORMAT(48H ---DO YOU WANT TO SELECT ANOTHER IDEAL SOLUTION,2X)
        END
C
C
C       SUBROUTINE ZMTOL IS THE MAIN SUBROUTINE FOR FLEXIBLE TOLERANCE
C       METHOD WHICH SOLVES A SINGLE CRITERION OPTIMIZATION PROBLEM
C
        SUBROUTINE ZMTOL
        COMMON/1/NX,NC,NIC,X(20),XAC(20),XK(20),XP(20),ID,IDR,R(40)
        COMMON/2/CONVER,SIZE,STEP,ALFA,BETA,GAMA,IN,INF,FDIFER,SEQL,K1,
       1K2,K3,K4,K5,K6,K7,K8,K9,X1(20,20),X2(20,20),SUM(20),SR(20),F(20),
       2FOLD(20),SCALE,FOLD,LFEAS,L5,L6,L7,L8,L9,R1A,R2A,R3A
        IF(ID.NE.1) GO TO 262
        DO 261 J=1,NX
    261 X(J)=XAC(J)
    262 ALFA = 1.0
        BETA = 0.5
        GAMA = 2.
        STEP = SIZE
        K1 = NX + 1
        K2 = NX + 2
        K3 = NX + 3
        K4 = NX + 4
        K5 = NX + 5
        K6 = NC + NIC
        K7 = NC + 1
        K8 = NC + NIC
        K9 = K8 + 1
        N = NX - NC
        N1 = N + 1
        IF(N1.GE.3) GO TO 50
        N1 = 3
        N = 2
     50 N2 = N + 2
        N3 = N + 3
```

```
      N4 = N + 4
      XN = N
      XNX = NX
      XN1 = N1
      R1A = 0.5*(SQRT(5.) - 1.)
      R2A = R1A*R1A
      R3A = R2A*R1A
      L5 = NX + 5
      L6 = NX + 6
      L7 = NX + 7
      L8 = NX + 8
      L9 = NX + 9
      ICONT = 1
      NCONT = 1
      PRINT 115
      PRINT 116, (X(J), J = 1, NX)
      FDIFER = 2.*(NC + 1)*STEP
      FOLD = FDIFER
      IN = N1
      CALL SUMR
      SR(N1) = SQRT(SEQL)
      IF(IDR.EQ.0) GO TO 501
      PRINT 763, FDIFER, SR(N1)
  501 IF(SR(N1).LT.FDIFER) GO TO 341
      CALL WRITEX
      PRINT 757
      INF = N1
      STEP = 0.05*FDIFER
      CALL FEASBL
      PRINT 764
      PRINT 116, (X2(INF,J),J = 1, NX)
      PRINT 765, SR(INF)
      IF(FOLD.LT.1.0E-09) GO TO 80
  341 IF(IDR.EQ.0) GOTO 237
      PRINT 35
      PRINT 758, ICONT, FDIFER
      CALL WRITEX
  237 STEP1 = STEP*(SQRT(XNX + 1.) + XNX- 1.)/(XNX*SQRT(2.))
      STEP2 = STEP*(SQRT(XNX + 1.) - 1.)/(XNX*SQRT(2.))
      ETA = (STEP1 + (XNX - 1.)*STEP2)/(XNX + 1.)
      DO 4 J = 1, NX
      X(J) = X(J) - ETA
    4 CONTINUE
      CALL START
      DO 9 I = 1, N1
      DO 9 J = 1, NX
      X2(I,J) = X1(I,J)
    9 CONTINUE
      DO 5 I = 1, N1
      IN = I
      DO 6 J = 1,NX
    6 X(J) = X2(I,J)
      CALL SUMR
      SR(I) = SQRT(SEQL)
      IF(SR(I).LT.FDIFER) GO TO 8
      INF=I
      CALL FEASBL
      IF(FOLD.LT.1.0E-09) GO TO 80
    8 CALL FCEL(X,FC)
      F(I)=FC
    5 CONTINUE
 1000 STEP = 0.05*FDIFER
      ICONT = ICONT + 1
      FH = F(1)
```

```
      LHIGH = 1
      DO 16 I = 2, N1
      IF(F(I).LT.FH) GO TO 16
      FH = F(I)
      LHIGH = I
   16 CONTINUE
   41 FL = F(1)
      LOW = 1
      DO 17 I = 2, N1
      IF(FL.LT.F(I)) GO TO 17
      FL = F(I)
      LOW = I
   17 CONTINUE
      DO 86 J = 1, NX
   86 X(J) = X2(LOW,J)
      IN = LOW
      CALL SUMR
      SR(LOW) = SQRT(SEQL)
      IF(SR(LOW).LT.FDIFER) GO TO 87
      INF = LOW
      CALL FEASBL
      IF(FOLD.LT.1.0E-09) GO TO 80
      CALL FCEL(X,FC)
      F(LOW)=FC
      GO TO 41
   87 CONTINUE
      DO 19 J = 1, NX
      SUM2 = 0.
      DO 20 I = 1, N1
   20 SUM2 = SUM2 + X2(I,J)
   19 X2(N2,J) = 1./XN*(SUM2-X2(LHIGH,J))
      SUM2 = 0.
      DO 36 I = 1, N1
      DO 36 J = 1, NX
      SUM2 = SUM2 + (X2(I,J) - X2(N2,J))**2
   36 CONTINUE
      FDIFER = (NC + 1)/XN1*SQRT(SUM2)
      IF(FDIFER.LT.FOLD) GO TO 98
      FDIFER = FOLD
      GO TO 198
   98 FOLD = FDIFER
  198 CONTINUE
      NCONT = NCONT + 1
      IF(NCONT.LT.4*N1) GO TO 37
      IF(ICONT.LT.1500) GO TO 337
      FOLD = 0.5*FOLD
  337 NCONT = 0
      IF(IDR.EQ.0) GO TO 37
      PRINT 35
      PRINT 758, ICONT,FDIFER
      CALL WRITEX
   37 IF(FDIFER.LT.CONVER) GO TO 31
      IF(LHIGH.EQ.1) GO TO 43
      FS = F(1)
      LSEC = 1
      GO TO 44
   43 FS = F(2)
      LSEC = 2
   44 DO 18 I = 1, N1
      IF(LHIGH.EQ.I) GO TO 18
      IF(F(I).LT.FS) GO TO 18
      FS = F(I)
      LSEC = I
   18 CONTINUE
```

```
      DO 61 J = 1, NX
      X2(N3,J) = X2(N2,J) + ALFA*(X2(N2,J) - X2(LHIGH,J))
   61 X(J) = X2(N3,J)
      IN = N3
      CALL SUMR
      SR(N3) = SQRT(SEQL)
      IF(SR(N3).LT.FDIFER) GO TO 82
      INF = N3
      CALL FEASBL
      IF(FOLD.LT.1.0E-09) GO TO 80
   82 CALL FCEL(X,FC)
      F(N3)=FC
      IF(F(N3).LT.F(LOW)) GO TO 84
      IF(F(N3).LT.F(LSEC)) GO TO 92
      GO TO 60
   92 DO 93 J = 1, NX
   93 X2(LHIGH,J) = X2(N3,J)
      SR(LHIGH) = SR(N3)
      F(LHIGH) = F(N3)
      GO TO 1000
   84 DO 23 J = 1, NX
      X2(N4,J) = X2(N3,J) + GAMA*(X2( N3,J) - X2(N2,J))
   23 X(J) = X2(N4,J)
      IN = N4
      CALL SUMR
      SR(N4) = SQRT(SEQL)
      IF(SR(N4).LT.FDIFER) GO TO 25
      INF = N4
      CALL FEASBL
      IF(FOLD.LT.1.0E-09) GO TO 80
   25 CALL FCEL(X,FC)
      F(N4)=FC
      IF(F(LOW).LT.F(N4)) GO TO 92
      DO 26 J = 1, NX
   26 X2(LHIGH,J) = X2(N4,J)
      F(LHIGH) = F(N4)
      SR(LHIGH) = SR(N4)
      GO TO 1000
   60 IF(F(N3).GT.F(LHIGH)) GO TO 64
      DO 65 J = 1, NX
   65 X2(LHIGH,J) = X2(N3,J)
   64 DO 66 J = 1,NX
      X2(N4,J) = BETA*X2(LHIGH,J) + (1. - BETA)*X2(N2,J)
   66 X(J) = X2(N4,J)
      IN = N4
      CALL SUMR
      SR(N4) = SQRT(SEQL)
      IF(SR(N4).LT.FDIFER) GO TO 67
      INF = N4
      CALL FEASBL
      IF(FOLD.LT.1.0E-09) GO TO 80
   67 CALL FCEL(X,FC)
      F(N4)=FC
      IF(F(LHIGH).GT.F(N4)) GO TO 68
      DO 69 J = 1, NX
      DO 69 I = 1, N1
   69 X2(I,J) = 0.5*(X2(I,J) + X2(LOW,J))
      DO 70 I = 1, N1
      DO 71 J = 1, NX
   71 X(J) = X2(I,J)
      IN = I
      CALL SUMR
      SR(I) = SQRT(SEQL)
      IF(SR(I).LT.FDIFER) GO TO 72
```

Appendix B

```
          INF = I
          CALL FEASBL
          IF(FOLD.LT.1.0E-09) GO TO 80
   72 CALL FCEL(X,FC)
   70 F(I)=FC
          GO TO 1000
   68 DO 73 J = 1, NX
   73 X2(LHIGH,J) = X2(N4,J)
          SR(LHIGH) = SR(N4)
          F(LHIGH) = F(N4)
          GO TO 1000
   81 IF(IDR.EQ.0) GO TO 502
          PRINT 760,ICONT,FDIFER
  502 CALL WRITEX
          GO TO 9999
   80 PRINT 760, ICONT, FDIFER
          CALL WRITEX
 9999 RETURN
   35 FORMAT(20(2H *))
  115 FORMAT(24H STARTING VALUES OF X(I))
  116 FORMAT(4E16.6)
  757 FORMAT(56H THE INITIAL X VECTOR DOES NOT SATISFY THE INITIAL TOLER
      1  14HANCE CRITERION)
  758 FORMAT(27H STAGE CALCULATION NUMBER =,I5,2X,
      1  26H THE TOLERANCE CRITERION =,E14.6)
  760 FORMAT(38H TOTAL NUMBER OF STAGES CALCULATIONS =,I5/
      1  24H THE CONVERGENCE LIMIT =,E14.6)
  763 FORMAT(10X,31HINITIAL TOLERANCE CRITERION   ,E12.5/
      1  10X,31HSUM OF VIOLATED CONSTRAINTS   ,E12.5)
  764 FORMAT(56H VECTOR X FOUND BY THE METHOD WHICH SATISFIES THE INITIA
      1  14HL TOLERANCE IS)
  765 FORMAT(30H SUM OF VIOLATED CONSTRAINTS =,E17.7)
          END
C
C
C         SUBROUTINE FEASBL MINIMIZES THE SUM OF SQUARE VALUES OF
C         THE VIOLATED CONSTRAINTS
C
          SUBROUTINE FEASBL
          DIMENSION R1(40),R2(40),R3(40),FLG(10),H(20)
          COMMON/1/NX,NC,NIC,X(20),XAC(20),XK(20),XP(20),ID,IDR,R(40)
          COMMON/2/CONVER,SIZE,STEP,DUM1,DUM2,DUM3,IN,INF,FDIFER,SEQL,K1,
         1K2,K3,K4,K5,K6,K7,K8,K9,X1(20,20),X2(20,20),SUM(20),SR(20),F(20),
         2ROLD(20),SCALE,FOLD,LFEAS,L5,L6,L7,L8,L9,R1A,R2A,R3A
          XNX = NX
          ICONT = 0
          ICHEK = 0
   25 CALL START
          DO 3 I = 1, K1
          DO 4 J = 1, NX
    4 X(J) = X1(I,J)
          IN = I
          CALL SUMR
    3 CONTINUE
   28 SUMH = SUM(1)
          INDEX = 1
          DO 7 I = 2, K1
          IF(SUM(I).LE.SUMH) GO TO 7
          SUMH = SUM(I)
          INDEX = I
    7 CONTINUE
          SUML = SUM(1)
          KOUNT = 1
```

```
      DO 8 I = 2, K1
      IF(SUML.LE.SUM(I)) GO TO 8
      SUML = SUM(I)
      KOUNT = I
    8 CONTINUE
      DO 9 J = 1, NX
      SUM2 = 0.
      DO 10 I = 1, K1
   10 SUM2 = SUM2 + X1(I,J)
      X1(K2,J) =1./XNX*(SUM2 - X1(INDEX,J))
      X1(K3,J) = 2.*X1(K2,J) - X1(INDEX,J)
    9 X(J) = X1(K3,J)
      IN = K3
      CALL SUMR
      IF(SUM(K3).LT.SUML) GO TO 11
      IF(INDEX.EQ.1) GO TO 38
      SUMS = SUM(1)
      GO TO 39
   38 SUMS = SUM(2)
   39 DO 12 I = 1, K1
      IF((INDEX - I).EQ.0) GO TO 12
      IF(SUM(I).LE.SUMS) GO TO 12
      SUMS = SUM(I)
   12 CONTINUE
      IF(SUM(K3).GT.SUMS) GO TO 13
      GO TO 14
   11 DO 15 J = 1, NX
      X1(K4,J) = X1(K2,J) + 2.*(X1(K3,J) - X1(K2,J))
   15 X(J) = X1(K4,J)
      IN = K4
      CALL SUMR
      IF(SUM(K4).LT.SUML) GO TO 16
      GO TO 14
   13 IF(SUM(K3).GT.SUMH) GO TO 17
      DO 18 J = 1, NX
   18 X1(INDEX,J) = X1(K3,J)
   17 DO 19 J = 1, NX
      X1(K4,J) = 0.5*X1(INDEX,J) + 0.5*X1(K2,J)
   19 X(J) = X1(K4,J)
      IN = K4
      CALL SUMR
      IF(SUMH.GT.SUM(K4)) GO TO 6
      DO 20 J = 1, NX
      DO 20 I = 1, K1
   20 X1(I,J) = 0.5*(X1(I,J) + X1(KOUNT,J))
      DO 29 I = 1, K1
      DO 30 J = 1, NX
   30 X(J) = X1(I,J)
      IN = I
      CALL SUMR
   29 CONTINUE
    5 SUML = SUM(1)
      KOUNT = 1
      DO 23 I = 2, K1
      IF(SUML.LT.SUM(I)) GO TO 23
      SUML = SUM(I)
      KOUNT = I
   23 CONTINUE
      SR(INF) = SQRT(SUM(KOUNT))
      DO 27 J = 1, NX
   27 X(J) = X1(KOUNT,J)
      GO TO 26
    6 DO 31 J = 1, NX
```

```
31 X1(INDEX,J) = X1(K4,J)
   SUM(INDEX) = SUM(K4)
   GO TO 5
16 DO 21 J = 1, NX
   X1(INDEX,J) = X1(K4,J)
21 X(J) = X1(INDEX,J)
   SUM(INDEX) = SUM(K4)
   SR(INF) = SQRT(SUM(K4))
   GO TO 26
14 DO 22 J = 1, NX
   X1(INDEX,J) = X1(K3,J)
22 X(J) = X1(INDEX,J)
   SUM(INDEX) = SUM(K3)
   SR(INF) = SQRT(SUM(K3))
26 ICONT = ICONT + 1
   DO 36 J = 1,NX
36 X2(INF,J) = X(J)
   IF(ICONT.LT.2*K1) GO TO 50
   ICONT = 0
   DO 24 J = 1, NX
24 X(J) = X1(K2,J)
   IN = K2
   CALL SUMR
   DIFER = 0.
   DO 57 I = 1, K1
57 DIFER = DIFER + (SUM(I) - SUM(K2))**2
   DIFER = 1./(K7*XNX)*SQRT(DIFER)
   IF(DIFER.GT.1.0E-14) GO TO 50
   IN=K1
   STEP = 20.*FDIFER
   CALL SUMR
   SR(INF) = SQRT(SEQL)
   DO 52 J = 1, NX
52 X1(K1,J) = X(J)
   DO 53 J = 1, NX
   FACTOR = 1.
   X(J) = X1(K1,J) + FACTOR*STEP
   X1(L9,J) = X(J)
   IN = L9
   CALL SUMR
   X(J) = X1(K1,J) - FACTOR*STEP
   X1(L5,J) = X(J)
   IN = L5
   CALL SUMR
56 IF(SUM(L9).LT.SUM(K1)) GO TO 54
   IF(SUM(L5).LT.SUM(K1)) GO TO 55
   GO TO 97
54 X1(L5,J) = X1(K1,J)
   SUM(L5) = SUM(K1)
   X1(K1,J) = X1(L9,J)
   SUM(K1) = SUM(L9)
   FACTOR = FACTOR + 1.
   X(J) = X1(K1,J) + FACTOR*STEP
   IN = L9
   CALL SUMR
   GO TO 56
55 X1(L9,J) = X1(K1,J)
   SUM(L9) = SUM(K1)
   X1(K1,J) = X1(L5,J)
   SUM(K1) = SUM(L5)
   FACTOR = FACTOR + 1.
   X(J) = X1(K1,J) - FACTOR*STEP
   IN = L5
```

```
       CALL SUMR
       GO TO 56
   97  H(J) = X1(L9,J) - X1(L5,J)
       X1(L6,J) = X1(L5,J) + H(J)*R1A
       X(J) = X1(L6,J)
       TN = L6
       CALL SUMR
       X1(L7,J) = X1(L5,J) + H(J)*R2A
       X(J) = X1(L7,J)
       IN = L7
       CALL SUMR
       IF(SUM(L6).GT.SUM(L7)) GO TO 68
       X1(L8,J) = X1(L5,J) + (1. - R3A)*H(J)
       X1(L5,J) = X1(L7,J)
       X(J) = X1(L8,J)
       IN = L8
       CALL SUMR
       IF(SUM(L8).GT.SUM(L6)) GO TO 76
       X1(L5,J) = X1(L6,J)
       SUM(L5) = SUM(L6)
       GO TO 75
   76  X1(L9,J) = X1(L8,J)
       SUM(L9) = SUM(L8)
       GO TO 75
   68  X1(L9,J) = X1(L6,J)
       X1(L8,J) = X1(L5,J) + R3A*H(J)
       X(J) = X1(L8,J)
       TN = L8
       CALL SUMR
       STEP = SIZE
       SUM(L9) = SUM(L6)
       IF(SUM(L7).GT.SUM(L8)) GO TO 71
       X1(L5,J) = X1(L8,J)
       SUM(L5) = SUM(L8)
       GO TO 75
   71  X1(L9,J) = X1(L7,J)
       SUM(L9) = SUM(L7)
   75  IF(ABS(X1(L9,J) - X1(L5,J)).GT.0.01*FDIFER) GO TO 97
       X1(K1,J) = X1(L7,J)
       X(J) = X1(L7,J)
       SUM(K1) = SUM(L5)
       SR(INF) = SQRT(SUM(K1))
       IF(SR(INF).LT.FDIFER) GO TO 760
   53  CONTINUE
       ICHEK = ICHEK + 1
       STEP = FDIFER
       IF(ICHEK.LE.2) GO TO 25
       FOLD = 1.0E-12
       PRINT 853
       PRINT 850
       PRINT 851, (X(J),J=1,NX)
       PRINT 852, FDIFER, SR(INF)
       GO TO 46
  760  DO 761 J = 1, NX
       X2(INF,J) = X1(K1,J)
  761  X(J) = X1(K1,J)
   50  IF(SR(INF).GT.FDIFER) GO TO 28
       IF(SR(INF).GT.0.) GO TO 35
       CALL FCEL(X,FINT)
       IF(NIC.EQ.0) GO TO 35
       DO 139 J = 1, NX
  139  X(J) = X2(INF,J)
       CALL OGRN(X,R)
```

```
         DO 40 J=1,NIC
         I=K7+J-1
  40 R1(I)=R(J)
         DO 41 J = 1, NX
  41 X(J) = X1(KOUNT,J)
         CALL OGRN(X,R)
         DO 42 J=1,NIC
         I=K7+J-1
  42 R3(I)=R(J)
         DO 43 J = 1, NX
         H(J) = X1(KOUNT,J) - X2(INF,J)
  43 X(J) = X2(INF,J) + 0.5*H(J)
         CALL OGRN(X,R)
         FLG(1) = 0.
         FLG(2) = 0.
         FLG(3) = 0.
         DO44 I=1,NIC
         J=K7+I-1
         IF(R3(J).GE.0.) GO TO 44
         FLG(1) = FLG(1) + R1(J)*R1(J)
         FLG(2)=FLG(2)+R(I)*R(I)
         FLG(3) = FLG(3) + R3(J)*R3(J)
  44 CONTINUE
         SR(INF) = SQRT(FLG(1))
         IF(SR(INF).LT.FDIFER) GO TO 35
         ALFA1 = FLG(1) - 2.*FLG(2) + FLG(3)
         BETA1 = 3.*FLG(1) - 4.*FLG(2) + FLG(3)
         RATIO = BETA1/(4.*ALFA1)
         DO 45 J = 1, NX
  45 X(J) = X2(INF,J) + H(J)*RATIO
         IN = INF
         CALL SUMR
         SR(INF) = SQRT(SEQL)
         IF(SR(INF).LT.FDIFER) GO TO 465
         DO 49 I = 1, 20
         DO 48 J = 1, NX
  48 X(J) = X(J) - 0.05*H(J)
         CALL SUMR
         SR(INF) = SQRT(SEQL)
         IF(SR(INF).LT.FDIFER) GO TO 465
  49 CONTINUE
 465 CALL FCEL(X,FC)
         IF(FINT.GT.FC) GO TO 46
         SR(INF) = 0.
         GO TO 35
  46 DO 47 J = 1, NX
  47 X2(INF,J) = X(J)
  35 CONTINUE
         DO 335 J = 1, NX
 335 X(J) = X2(INF,J)
 850 FORMAT(56H IT IS NOT POSSIBLE TO SATISFY THE VIOLATED CONSTRAINTS
    1 /59H FOR STARTING VALUES OF X(I).THE SEARCH WILL BE TERMINATED
    2 /42H PLEASE CHOOSE NEW STARTING VALUES OF X(I))
 851 FORMAT(56H VECTOR X(I) FOR WHICH THE CONSTRAINTS COULD NOT BE SATI
    1 8HSFIED IS)
 852 FORMAT(26H THE TOLERANCE CRITERION =,E14.6/
    1 45H THE SQUARE ROOT OF THE CONSTRAINTS SQUARED =,E16.6)
 853 FORMAT(48H ***FT METHOD FAILS TO FIND A FEASIBLE POINT*** )
         RETURN
         END
C
C
C        SUBROUTINE START FINDS THE STARTING VALUES OF X(I) WHICH SATISFY
```

```
C       THE INITIAL TOLERANCE
C
        SUBROUTINE START
        DIMENSION A(20,20)
        COMMON/1/NX,NC,NIC,X(20),XAC(20),XK(20),XP(20),ID,IDR,R(40)
        COMMON/2/CONVER,SIZE,STEP,ALFA,BETA,GAMA,IN,INF,FDIFER,SEQL,K1,
       1K2,K3,K4,K5,K6,K7,K8,K9,X1(20,20),X2(20,20),SUM(20),SR(20),F(20),
       2POLD(20),SCALE,FOLD,LFEAS,L5,L6,L7,L8,L9,R1A,R2A,R3A
        VN = NX
        STEP1 = STEP/(VN*SQRT(2.))*(SQRT(VN + 1.) + VN - 1.)
        STEP2 = STEP/(VN*SQRT(2.))*(SQRT(VN + 1.) - 1.)
        DO 1 J = 1, NX
      1 A(1,J) = 0.
        DO 2 I = 2, K1
        DO 4 J = 1, NX
      4 A(I,J) = STEP2
        L = I - 1
        A(I,L) = STEP1
      2 CONTINUE
        DO 3 I = 1, K1
        DO 3 J = 1, NX
      3 X1(I,J) = X(J) + A(I,J)
        RETURN
        END
C
C
C       SUBROUTINE SUMR CALCULATES THE SUM OF SQUARE VALUES OF THE
C       VIOLATED CONSTRAINTS IN ORDER TO BE COMPARED WITH THE TOLERANCE
C       CRITERION
C
        SUBROUTINE SUMR
        COMMON/1/NX,NC,NIC,X(20),XAC(20),XK(20),XP(20),ID,IDR,R(40)
        COMMON/2/CONVER,SIZE,STEP,ALFA,BETA,GAMA,IN,INF,FDIFER,SEQL,K1,
       1K2,K3,K4,K5,K6,K7,K8,K9,X1(20,20),X2(20,20),SUM(20),SR(20),F(20),
       2POLD(20),SCALE,FOLD,LFEAS,L5,L6,L7,L8,L9,R1A,R2A,R3A
        SUM(IN) = 0.
        SEQL = 0.
        IF(NIC.EQ.0) GO TO 4
        CALL OGRN(X,R)
        DO 1 J=1,NIC
        IF(R(J).GE.0.) GO TO 1
        SEQL = SEQL + R(J)*R(J)
      1 CONTINUE
      4 IF(NC.EQ.0) GO TO 3
        CALL OGRR(X,R)
        DO 2 J = 1, NC
      2 SEQL = SEQL + R(J)*R(J)
      3 SUM(IN) = SEQL
        RETURN
        END
C
C
C       SUBROUTINE SEKLOS IS THE MAIN SUBROUTINE FOR THE DIRECT AND RANDOM
C       SEARCH METHOD WHICH SOLVES A SINGLE CRITERION OPTIMIZATION
C       PROBLEM
C
        SUBROUTINE SEKLOS
        COMMON/1/NX,NC,NIC,X(20),XAC(20),XK(20),XP(20),ID,IDR,R(40)
        COMMON/3/MAXM,NVIOL,FPK,WK,NNDEX,FEX,KRK,NTEST,LW,XV(20),NH,F,
       6NTMC,NRAZ,XSTRT(20),A1(20),A2(20),A3(20),A4(20)
        DIMENSION XEX(20)
        IF(ID.NE.1) GO TO 262
        DO 261 I=1,NX
```

```
261 XSTRT(I)=XAC(I)
    GO TO 263
262 DO 264 I=1,NX
264 XSTRT(I)=X(I)
263 PRINT 22
    PRINT 23,(XSTRT(I),I=1,NX)
    LW=0
    KL=1
340 CONTINUE
    DO 100 I=1,NX
    X(I)=0.0
    A1(I)=0.0
    A2(I)=0.0
    A3(I)=0.0
100 A4(I)=0.0
    NNDEX=1
    KOUNT=0
  6 CALL SZUK
    CALL LOS
    IF(KRK.EQ.1) GO TO 4
    GO TO 16
  4 IF(IDR.EQ.0) GO TO 18
    PRINT 29
    PRINT 24,KOUNT
    PRINT 25,F,(X(I),I=1,NX)
 18 KOUNT=KOUNT+1
    IF(KOUNT.LE.NH) GO TO 13
    GO TO 16
 13 DO 14 I=1,NX
 14 XSTRT(I)=X(I)
    GO TO 6
 16 IF(IDR.EQ.0) GO TO 19
    PRINT 26,KL
    CALL WRITEX
 19 IF(KL.NE.1) GO TO 510
    FEX=F
    DO 512 I=1,NX
512 XEX(I)=X(I)
    GO TO 514
510 IF(F.GE.FEX) GO TO 514
    FEX=F
    DO 513 I=1,NX
513 XEX(I)=X(I)
514 CONTINUE
    IF(KL.GT.NPAZ) GO TO 600
    KL=KL+1
    CALL LOSMC
    DO 400 I=1,NX
400 XSTRT(I)=XV(I)
    IF(IDR.EQ.0) GO TO 500
    KZ=KL-1
    PRINT 27,KZ
    PRINT 28,(XV(I),I=1,NX)
500 GO TO 340
600 F=FEX
    DO 300 I=1,NX
300 X(I)=XEX(I)
    CALL WRITEX
    RETURN
 22 FORMAT(24H STARTING VALUES OF X(I))
 23 FORMAT(4E16.6)
 29 FORMAT(/45H A SATISFYING RESULT OF SHOTGUN SEARCH IN LOS )
 24 FORMAT(18H THE SEARCH NUMBER,I4)
```

```
  25 FORMAT(5X,4HF = ,E17.7,//,5X,15H VECTOR OF X(I),/(4X,4E17.7),//,1X,1
     1 9(1H-))
  26 FORMAT(10(2H *)/,14H RESULT NUMBER,I5,24H OF DIRECT SEARCH METHOD)
  27 FORMAT(10(2H *)/,14H RESULT NUMBER,I5,24H OF RANDOM SEARCH METHOD)
  28 FORMAT(25H GENERATED VECTOR OF X(I),/(4E17.7))
     END
C
C
C     SUBROUTINE SZUK REALIZES THE HOOK AND JEFVES DIRECT SEARCH METHOD
C
      SUBROUTINE SZUK
      COMMON/1/NX,NC,NIC,X(20),XAC(20),XK(20),XP(20),ID,IDR,R(40)
      COMMON/3/MAXM,NVIOL,FPK,WK,NNDEX,FEX,KRK,NTEST,LW,XV(20),NH,F,
     7NTMC,NRAZ,XSTRT(20),XO(20),XB(20),DXXX(20),TXXX(20)
      NVIOL1=1
      KKK=0
      M1=0
  20  K1=1
      K2=NX
  30  DO 40 I=K1,K2
      DXXX(I)=0.0
      TXXX(I)=0.0
      XO(I)=0.0
  40  XB(I)=0.0
      DO 60 I=K1,K2
  60  X(I)=XSTRT(I)
      DO 70 I=K1,K2
  70  XO(I)=X(I)
      DO 80 I=K1,K2
      DXXX(I)=FPK*(XK(I)-XP(I))
  80  TXXX(I)=DXXX(I)*WK
      NCALL=1
 100  CONTINUE
 101  CALL KARA(X,UART,NIC,NC,NVIOL)
 110  IF(NCALL.NE.1) GO TO 120
      UARTO=UART
 120  CONTINUE
      IF(NVIOL.EQ.0)NVIOL1=0
      IF(NNDEX.EQ.1) GO TO 130
      IF(NVIOL1.EQ.0) GO TO 385
 130  GO TO (170,200,210,355) NCALL
 170  CONTINUE
 180  NFAIL=0
      I=K1-1
 270  I=I+1
      X(I)=X(I)+DXXX(I)
      NCALL=2
      GO TO 100
 200  CONTINUE
      IF(UART.LT.UARTO) GO TO 230
      X(I)=X(I)-2.0*DXXX(I)
      NCALL=3
      GO TO 100
 210  CONTINUE
      IF(UART.LT.UARTO) GO TO 230
      NFAIL=NFAIL+1
      X(I)=X(I)+DXXX(I)
      GO TO 240
 230  UARTO=UART
 240  IF(I.LT.K2) GO TO 270
 250  IF(NFAIL.EQ.NX) GO TO 260
      GO TO 315
 260  DO 280 I=K1,K2
```

```
      IF(DXXX(I).GT.TXXX(I)) GO TO 290
280 CONTINUE
      GO TO 385
290 DO 310 I=K1,K2
310 DXXX(I)=DXXX(I)/2.0
      GO TO 180
315 DO 320 I=K1,K2
320 XB(I)=X(I)
      M1=M1+1
      IF(NNDEX.EQ.1) GO TO 330
      GO TO 340
330 KKK=KKK+1
      IF(KKK.NE.IDR) GO TO 340
      CALL FCEL(X,FZ)
      PRINT 5,M1
      PRINT 6,FZ,(X(I),I=1,NX)
357 KKK=0
340 CONTINUE
      IF(M1.GT.MAXM)GO TO 385
      DO 350 I=K1,K2
350 X(I)=X(I)+(X(I)-XO(I))
      NCALL=4
      GO TO 100
355 CONTINUE
      IF(UART.LT.UARTO) GO TO 370
      DO 360 I=K1,K2
      XO(I)=XB(I)
360 X(I)=XB(I)
      GO TO 180
370 DO 380 I=K1,K2
380 XO(I)=XB(I)
      UARTO=UART
      GO TO 180
385 CALL FCEL(X,F)
103 CALL KARA(X,UART,NIC,NC,NVIOL)
105 IF(NVIOL.EQ.0) GO TO 387
      IF(IDR.EQ.0) GO TO 387
      IF(M1.GT.MAXM) PRINT 4,MAXM
387 RETURN
  5 FORMAT(10H RESULT OF,I5,31H CYCLE OF FUNCTION MINIMIZATION )
  6 FORMAT(6H FZ = ,E17.7,/,15H VECTOR OF X(I),/,(4E16.6))
  4 FORMAT(42H ***PROGRAM FAILS TO FIND A FEASIBLE POINT,/,6H MAXM=,I4
  1   ,)
      END
C
C
C      SUBROUTINE LOS GENERATES A BETTER STARTING POINT FOR DIRECT
C      SEARCH METHOD
C
      SUBROUTINE LOS
      COMMON/1/NX,NC,NIC,X(20),XAC(20),XK(20),XP(20),ID,IDR,R(40)
      COMMON/3/MAXM,NVIOL,FPK,WK,NNDEX,FEX,KRK,NTEST,LW,XV(20),NH,F,
     8NTMC,NRAZ,XSTRT(20),RR(20),XX(20),RF(20),BLO(20)
      IF(LW.EQ.1) GO TO 9
      CALL RANSET(7)
      LW=1
  9 CONTINUE
      UMIN=F
      KRK=0
      DO 1 I=1,NX
  1 RF(I)=10.0*FPK*ABS(XK(I)-XP(I))
      DO 4 J=1,NTEST
      DO 8 I=1,NX
```

```
8     RR(I)=RANF(RP)
      DO 2 I=1,NX
      IF(X(I)-RF(I).LT.0) GO TO 10
      XX(I)=(X(I)-RF(I))+RR(I)*2.0*RF(I)
      GO TO 2
  10  XX(I)=XP(I)+RR(I)*2.0*RF(I)
   2  CONTINUE
      CALL KARA(XX,UTEST,NIC,NC,NVIOL)
      IF(UTEST.GE.UMIN) GO TO 4
      IF(NVIOL.NE.0) GO TO 4
      UMIN=UTEST
      F=UTEST
      DO 3 I=1,NX
   3  Y(I)=XX(I)
      KRK=1
   4  CONTINUE
      RETURN
      END
C
C
C     SUBROUTINE KARA CALCULATES A VALUE OF AN ARTIFICIAL OBJECTIVE
C     FUNCTION WHICH PENALIZES NONFEASIBLE SOLUTIONS
C
      SUBROUTINE KARA(X,UART,NIC,NC,NVIOL)
      DIMENSION X(20),H(20),G(20)
      ZERO=-1.0E-10
      NVIOL=0
      SUM1=0.0
      SUM2=0.0
      CALL FCEL(X,F)
      IF(NIC.EQ.0) GO TO 2
      CALL OGRN(X,G)
      DO 1 I=1,NIC
      IF(G(I).GE.ZERO) GO TO 1
      SUM1=SUM1+ABS(G(I)*10.E+20)
      NVIOL=NVIOL+1
   1  CONTINUE
   2  IF(NC.EQ.0) GO TO 115
      CALL OGRR(X,H)
      DO 3 I=1,NC
   3  SUM2=SUM2+ABS(H(I))*10.0E+20
 115  UART=F+SUM1+SUM2
      RETURN
      END
C
C
C     SUBROUTINE LOSMC TRAYS TO GENERATE A BETTER SOLUTION IN THE
C     VICINITY OF THE MINIMUM FOUND BY DIRECT SEARCH METHOD
C
      SUBROUTINE LOSMC
      DIMENSION XMC(20),RO(20)
      COMMON/1/NX,NC,NIC,X(20),XAC(20),XK(20),XP(20),ID,IDR
      COMMON/3/MAXM,NVIOL,FPK,WK,NNDFX,FEX,KRK,NTEST,LW,XV(20),NH,F,
     6NTMC,NRAZ,XSTRT(20),A1(20),A2(20),A3(20),A4(20)
      CALL RANSET(9)
      DO 90 I=1,NX
  90  XV(I)=0.0
      IRA=0
      DO 70 J=1,NTMC
      DO 80 I=1,NX
  80  RO(I)=RANF(R)
      DO 30 I=1,NX
  30  XMC(I)=XP(I)+RO(I)*(XK(I)-XP(I))
```

```
      CALL KARA(XMC,UTMC,NIC,NC,NVIOL)
      IF(NVIOL.NE.0) GO TO 70
      IF(IRA.EQ.1) GO TO 200
      UMIN=UTMC
      DO 210 I=1,NX
210   XV(I)=XMC(I)
      IRA=1
200   CONTINUE
      IF(UTMC.GE.UMIN) GO TO 70
      UMIN=UTMC
      DO 40 I=1,NX
40    XV(I)=XMC(I)
70    CONTINUE
      IF(IRA.EQ.1) GO TO 150
      PRINT 44
150   CONTINUE
      RETURN
44    FORMAT(56H ***NO FEASIBLE SOLUTION HAS BEEN GENERATED BY RANDOM SE
     1 11HARCH METHOD)
      END
C
C
C
C     SUBROUTINE WRITEX PRINTS THE RESULTS OF CALCULATIONS
C
      SUBROUTINE WRITEX
      COMMON/1/NX,NC,NIC,X(20),XAC(20),XK(20),XP(20),ID,IDR,R(40)
      COMMON/2/CONVER,SIZE,STEP,ALFA,BETA,GAMA,IN,INF,FDIFER,SEQL,K1,
     1K2,K3,K4,K5,K6,K7,K8,K9,X1(20,20),X2(20,20),SUM(20),SR(20),F(20),
     2ROLD(20),SCALE,FOLD,LFEAS,L5,L6,L7,L8,L9,R1A,R2A,R3A
      COMMON/3/MAXM,NVIOL,FPK,WK,NNDEX,FEX,KRK,NTEST,LW,XV(20),NH,FCE,
     6NTMC,NRAZ,XSTRT(20),A1(20),A2(20),A3(20),A4(20)
      COMMON/C/M,A(20),P(20),N,KK,MIA(20),WSP(40),IS,IPO
      CALL FCEL(X,FC)
      KWP=MIA(IS)
      GO TO (21,21,21,22,22),KWP
21    A(M)=FC
      IF(KK.EQ.1) GO TO 11
      PRINT 1,(LI,P(LI),LI=1,N)
      GO TO 10
11    CONTINUE
      PRINT 7,FC
      MM=M
      M=M-N
      DO 100 I=1,N
      CALL FCEL(X,FC)
      PRINT 8,I,FC
      M=M+1
100   CONTINUE
      M=MM
      GO TO 10
22    PRINT 9,FC,(LI,P(LI),LI=1,N)
10    MIA(IS)=KWP
      PRINT 2, (Y(J), J = 1, NX)
      IF(NC.EQ.0) GO TO 6
      CALL OGRR(X,R)
      PRINT 3, (R(J), J = 1, NC)
6     IF(NIC.EQ.0) GO TO 5
      CALL OGPN(X,R)
      PRINT 4,(R(J),J=1,NIC)
1     FORMAT(9H VALUE OF,I3,21H OBJECTIVE FUNCTION =,E17.7)
7     FORMAT(35H VALUE OF THE MINIMIZED QUANTITY = ,E17.7)
8     FORMAT(9H VALUE OF,I3,21H OBJECTIVE FUNCTION =,E17.7)
9     FORMAT(34H VALUE OF THE MINIMIZED QUANTITY =,E17.7,//,(9H VALUE OF,
```

```
      1  I3,21H ORJICTIVE FUNCTION =,E17.7))
      2 FORMAT(29H VECTOR OF DECISION VARIABLES  ,/(4E17.7))
      3 FORMAT(27H EQUALITY CONSTRAINT VALUES ,/(4E17.7))
      4 FORMAT(29h INEQUALITY CONSTRAINT VALUES ,/(4E17.7))
      5 RETURN
        END
C
C
C     SUBROUTINE TABL CALCULATES AND PRINTS THF PAYOFF TABLES
C
        SUBROUTINE TABL(NTAB)
        COMMON/C/M,A(20),P(20),N,KK,MIA(20),WSP(40),IS,IPO
        COMMON/D/DW(20,20),DB(20,20),PO(20,20)
        IF(NTAB.EQ.1) GO TO 2
        J=M
        DO 1 I=1,N
      1 PO(I,J)=P(I)
        PETURN
      2 DO 3 J=1,N
        DO 3 I=1,N
        DB(I,J)=ABS(A(I)-PO(I,J))
        IF(PO(I,J).EQ.0.0) PO(I,J)=0.1E-50
        IF(A(I).EQ.0.0) A(I)=0.1E-50
      3 DW(I,J)=AMAX1(ABS(DB(I,J)/A(I)),ABS(DB(I,J)/PO(I,J)))
        PRINT 11
        DO 4 J=1,N
      4 PRINT 10,(DB(I,J),I=1,N)
        PRINT 12
        DO 5 J=1,N
      5 PRINT 10,(DW(I,J),I=1,N)
     10 FORMAT(8E16.7)
     11 FORMAT(1X,36(1H-),//,1X,36HPAYOFF TABLE FOR FUNCTION INCREMENTS,/,
      1  1X,36(1H-))
     12 FORMAT(1X,45(1H-)/
      1  46H PAYOFF TABLE FOR FUNCTION RELATIVE INCREMENTS /1X,45(1H-))
        RETURN
        END
C
C
C     SUBROUTINE MINMAX DETERMINES THE MINIMUM OF FUNCTION RELATIVE
C     INCREMENTS
C
        SUBROUTINE MINMAX(AMAX)
        COMMON/1/NX,NC,NIC,X(20),XAC(20),XK(20),XP(20),ID,IDR,R(40)
        COMMON/C/M,A(20),P(20),N,KK,MIA(20),WSP(40),IS,IPO
        MM=M
        M=M-N
        AMAX=-1.
        DO 1 I=1,N
        CALL ODCHYL(DH)
        IF(AMAX.LT.DH) AMAX=DH
        M=M+1
      1 CONTINUE
        M=MM
        RETURN
        END
C
C
C     SUBROUTINE ODCHYL CALCULATES THE FUNCTION RELATIVE INCREMENTS
C
        SUBROUTINE ODCHYL(DH)
        COMMON/1/NX,NC,NIC,X(20),XAC(20),XK(20),XP(20),ID,IDR,R(40)
        COMMON/C/M,A(20),P(20),N,KK,MIA(20),WSP(40),IS,IPO
```

```
        CALL FCEL1(X,F)
        KPP=MIA(IS)
        G=F-A(M)
        ZERO=0.1E-50
        IF(F.EQ.0.0) F=ZERO
        IF(A(M).EQ.0.0) A(M)=ZERO
        GO TO (31,32,33,34,34),KPP
    31  DH=AMAX1(ABS(G/A(M)),ABS(G/F))
        MIA(IS)=KPP
    32  RETURN
    33  DH=AMAX1(WSP(M)*ABS(G/A(M)),WSP(M)*ABS(G/F))
        MIA(IS)=KPP
    34  RETURN
        END
C
C
C
C       SUBROUTINE FCEL ORGANIZES CALCULATIONS FOR THE MINIMIZED QUANTITY
C
        SUBROUTINE FCEL(X,F)
        DIMENSION X(20)
        COMMON/C/M,A(20),P(20),N,KK,MIA(20),WSP(40),IS,IPO
        KNT=MIA(IS)
        GO TO (1,1,1,2,2),KNT
     1  KTP=N+1
        IF(KNT.EQ.2) GO TO 2
        IF(M.EQ.KTP) GO TO 11
     2  CALL FCEL1(X,F)
        MIA(IS)=KNT
        RETURN
    11  CALL MINMAX(F)
        MIA(IS)=KNT
        RETURN
        END
C
C
C
C       SUBROUTINE FCEL1 CALCULATES A VALUE OF THE MINIMIZED QUANTITY
C
        SUBROUTINE FCEL1(X,F)
        DIMENSION X(20)
        COMMON/C/M,A(20),P(20),N,KK,MIA(20),WSP(40),IS,IPO
        KUP=MIA(IS)
        CALL FCELU(X,P)
        GO TO (21,21,21,22,26),KUP
    21  KTR=N+1
        IF(M.EQ.KTR) GO TO 12
        F=P(M)
        MIA(IS)=KUP
        RETURN
    12  F=0.0
        DO 3 L=1,N
        IF(A(L).EQ.0.0) A(L)=0.1E-50
     3  F=F+(P(L)/A(L)-1.0)**IPO
        MIA(IS)=KUP
        RETURN
    22  F=0.0
        DO 4 L=1,N
     4  F=F+WSP(L)*P(L)
        MIA(IS)=KUP
        RETURN
    26  F=0.0
        DO 8 L=1,N
        IF(A(L).EQ.0.0) A(L)=0.1E-50
     8  F=F+WSP(L)*(P(L)/A(L))
```

```
MIA(IS)=KUP
RETURN
END
```

A FORTRAN program for network multicriterion optimization

PURPOSE

To find a set of Pareto optimal paths and the min–max optimal path using the method described in Section 5.3.

STRUCTURE OF THE PROGRAM

The program consists of the main routine and three subroutines MINMAX, PARETO and DRUK.

(1) The main routine seeks the shortest path for each objective function separately and suboptimal paths for the first objective function.

(2) Subroutine MINMAX selects the min–max optimal path from a set of paths. The dummy arguments correspond to the symbols from Fig. 5.6 as follows

$$K = k, \ M = l, \ LZ = l^*, \ HOP = \bar{f}^0, \ I2 = d^{(l)}, \ H2 = \bar{f}(d^{(l)}),$$
$$H3 = \bar{f}(d^*), \ IN3 = d^*, \ D3 = \bar{z}(d^*).$$

The remaining arguments are:

NNN = number of nodes for the path d^*,
IZ   = number of nodes for the path $d^{(l)}$,
IPP = if equals one then formula (5.13) is satisfied,
IPK = if equals one then v_1^* equals zero and there is no better path in the network (see Section 4.2).
ICA = accuracy of comparing the function relative increments (see Appendix A).

(3) Subroutine PARETO selects the set of Pareto optimal paths from a set of paths. The dummy arguments correspond to the symbols from Fig. 5.6 as follows

$$K = k, \ JA = j^a, \ IN2 = d^{(l)}, \ H2 = f(d^{(l)}), \ IN4 = d_j^p, \ IHP = f_j^p$$

The remaining arguments are:

NN = number of nodes for the path d_j^p,
IZ = number of nodes for the path $d^{(l)}$.

(4) Subroutine DRUK prints the results of calculations.

INPUT VARIABLES

First card

K   = number of objective functions,
N   = number of nodes in the network,
A   = an aribitrarily chosen integer value such that, for any path the value of any objective function is less than A,
ILA = number of suboptimal paths to be examined,

IPR = $\begin{cases} 0, \text{ only final results are printed,} \\ 1, \text{ all suboptimal paths for the first objective function are printed,} \end{cases}$

ICA = accuracy of comparing the function relative increments.

Second and the following cards

J3,((BN(I1,I,J),I1 = 1,K),J = I2,J3) where
BN(I1,I,J) = matrix of the network,
I1   = number of objective function,
I,J   = node numbers,
I2   = number of the node considered, where I2 = 1, 2, ..., N−1,
J3   = number of the last node associated with I2 node.

Nodes must be numbered in ascending order from the beginning of the network (node number 1) to the end of the network (node number N). This means that if the arc between nodes I and J is directed from the node I to the node J, then I < J. Values of the elements of the matrix BN(I1,I,J) must be integer and greater than zero. These elements correspond to the network assignment as follows: If there is a two-argument relation between nodes I and J (there is an arc between nodes I and J), the value of BN(I1,I,J) must be equal to the value of the I1th objective function associated with the arc between these nodes. Otherwise the value of BN(I1,I,J) must be equal to A. For example

$[5,65,220]^T$

$u_3, \quad u_5$

$$BN(1,3,5) = 5$$
$$BN(2,3,5) = 65$$
$$BN(3,3,5) = 220$$

OUTPUT VARIABLES

The program prints input data and the results with comments. After the program listing the output for Example 5.1 is given.

THE PROGRAM LISTING

```
      PROGRAM OPTWS (INPUT,OUTPUT,TAPE1=INPUT,TAPE3=OUTPUT)
C
C
C     PROGRAM OPTWS
C     NETWORK MULTICRITERION OPTIMIZATION
C     THIS ROUTINE SEEKS THE SHORTEST PATHS FOR EACH OBJECTIVE FUNCTION
C     SEPARATELY, AND SUBOPTIMAL PATHS FOR THE FIRST OBJECTIVE FUNCTION
C
      DIMENSION DH(5),D3(5),NN(100),IN4(30,100),IHP(5,100),IH(5),DW(5,5)
     1 ,BN(5,30,30),H(30,30),V(90,90),U(90,90),VO(90),E(90),T(90),
     2 BA(90),P(90),VA(90),UO(90),VP(90),IN(30),IN1(30),IN2(30),IN3(30)
     3 ,HOP(5),H2(5),H3(5),IB(5),IDH(5),ID(5),IDZ(5)
      INTEGER BN,H,V,U,VO,E,T,BA,P,VA,UO,VP,HOP,H2,H3,Z,TA,HN,HN1,A,B,
     1 BP,BE,ZP
      READ(1,100) K,N,A,ILA,IPR,ICA
      WRITE(3,101) K,N,A,ILA,IPR,ICA
      N1=N-1
      DO 202 I=1,N1
      I2=I+1
      READ(1,100) J3,((BN(I1,I,J),I1=1,K),J=I2,J3)
      J4=J3+1
      DO 202 J=J4,N
      DO 202 I1=1,K
  202 BN(I1,I,J)=A
      DO 1 I1=1,K
      WRITE(3,102) I1
      N1=N-1
      DO 1 I=1,N1
      I2=I+1
    1 WRITE(3,110) I,(BN(I1,I,J),J=I2,N)
      WRITE(3,114)
      IPP=0
      JA=1
      IPK=0
      DO 90 I1=1,K
      IHP(I1,1)=1000000
   90 IDZ(I1)=100*10**ICA
      I1=1
  200 DO 99 I=1,N
      DO 99 J=I,N
```

```
 99 H(I,J)=BN(I1,I,J)
208 J=1
 96 VO(J)=A
    V(1,J)=0
    T(J)=0
    J=J+1
    IF(J.LE.N) GO TO 98
    GO TO 2
 98 I=1
 95 IF(H(I,J).EQ.A) GO TO 97
    P(J)=I
    GO TO 96
 97 I=I+1
    IF(I.LT.J) GO TO 95
    GO TO 96
  2 V(1,1)=1
    V(2,1)=0
    TA=0
    M=1
 35 J=2
 17 E(1)=0
    RA(J)=1
    IF(M.EQ.1) GO TO 3
    IF(V(1,J).EQ.0) GO TO 4
  3 VO(J)=A
    I=P(J)
 16 RP=1
 13 R=RP
    IF(V(B,I).EQ.0) GO TO 5
    VA(J)=V(B,I)+H(I,J)
    RB=1
 10 R=BB
    IF(V(B,J).EQ.0) GO TO 6
    L=1
  9 IF(E(L).EQ.0) GO TO 7
    IF(E(L).EQ.B) GO TO 8
    L=L+1
    GO TO 9
  8 BB=BB+1
    GO TO 10
  7 IF(VA(J).EQ.V(B,J)) GO TO 81
    GO TO 8
 81 U(B,J)=I
    L=1
 12 IF(E(L).EQ.0) GO TO 11
    L=L+1
    GO TO 12
 11 E(L)=B
    E(L+1)=0
 14 BP=BP+1
    GO TO 13
  6 IF(VA(J).LT.VO(J)) GO TO 141
    GO TO 14
141 VO(J)=VA(J)
    UO(J)=I
    GO TO 14
  5 VA(J)=VO(I)+H(I,J)
    IF(VA(J).LT.VO(J)) GO TO 151
    GO TO 15
151 VO(J)=VA(J)
    UO(J)=I
 15 I=I+1
    IF(I.LT.J) GO TO 161
    GO TO 4
```

```
161 IF(H(I,J).EQ.A) GO TO 15
    GO TO 16
  4 J=J+1
    IF(J.LE.N) GO TO 17
    J=N
    B=1
 19 IF(V(B,J).EQ.0) GO TO 18
    B=B+1
    GO TO 19
 18 V(B,J)=V0(J)
    V(B+1,J)=0
    VP(J)=V0(J)
    HN=V0(J)-1
    I=J
 21 I=I-1
    IF(I.EQ.1) GO TO 20
    IF(I.NE.U0(J)) GO TO 21
    Z=1
 28 VP(I)=VP(J)-H(I,J)
    IN(Z)=I
    IF(V0(I).EQ.VP(I)) GO TO 22
 38 B=BA(I)
    IN(Z)=I
 24 IF(V(B,I).EQ.VP(I)) GO TO 23
    B=B+1
    GO TO 24
 22 B=1
 26 IF(V(B,I).EQ.0) GO TO 25
    B=B+1
    GO TO 26
 25 V(B,I)=VP(I)
    V(B+1,I)=0
    J=I
 27 I=I-1
    IF(I.EQ.1) GO TO 20
    IF(I.NE.U0(J)) GO TO 27
 30 Z=Z+1
    GO TO 28
 23 IF(V(B+1,I).EQ.VP(I)) GO TO 29
    T(I)=0
    J=I
 31 I=I-1
    IF(I.EQ.1) GO TO 20
    IF(I.EQ.U(B,J)) GO TO 30
    GO TO 31
 29 T(I)=1
    TA=1
    BA(I)=B+1
    J=I
 32 I=I-1
    IF(I.EQ.1) GO TO 20
    IF(I.EQ.U(B,J)) GO TO 30
    GO TO 32
 20 ZP=Z
    HN1=HN
    DO 33 Z=1,ZP
    IN2(Z+1)=IN(Z)
 33 IN1(Z)=IN(Z)
    IN2(1)=N
    IN2(ZP+2)=1
    IZ=ZP+2
    IF(I1.GT.K) GO TO 201
    HOP(I1)=HN
    WRITE(3,103) I1
    DO 352 IV=1,K
```

```
      IH(IV)=0
      DO 351 IM=2,IZ
      IM1=IN2(IM-1)
      IM2=IN2(IM)
  351 IH(IV)=IH(IV)+BN(IV,IM2,IM1)
  352 DW(I1,IV)=IH(IV)
      DH(K+1)=0.
      CALL DRUK(K,IZ,IH,IN2,DH)
      I1=I1+1
      IF(I1.LE.K) GO TO 200
      DH(K+1)=1.
      WRITE(3,105)
      DO 42 IQ=1,K
      DO 42 IV=1,K
      C2=HOP(IV)
   42 DW(IQ,IV)=(DW(IQ,IV)-C2)/C2
      DO 43 IV=1,K
   43 WRITE(3,107)(DW(IQ,IV),IQ=1,K)
      IF(IPR) 217,217,213
  213 WRITE(3,115)
  217 DO 203 I=1,N
      DO 203 J=I,N
  203 H(I,J)=BN(1,I,J)
      GO TO 208
  201 IF(HN.GE.A) GO TO 166
      I2=1
  206 H2(I2)=0
      J1=1
  204 J2=J1+1
      K1=IN2(J1)
      K2=IN2(J2)
      H2(I2)=H2(I2)+BN(I2,K2,K1)
      J1=J1+1
      IF(J1+1.LE.IZ) GO TO 204
      C1=H2(I2)
      C2=HOP(I2)
      DH(I2)=(C1-C2)/C2
  162 I2=I2+1
      IF(I2.LE.K) GO TO 206
      IF(IPR) 205,205,207
  207 WRITE(3,111) M
      CALL DRUK(K,IZ,H2,IN2,DH)
  205 CALL PARETO(K,NN,IZ,JA,IN2,H2,IN4,IHP)
      IF(JA.EQ.100) GO TO 167
      CALL MINMAX(K,NNM,IZ,M,LZ,HOP,IN2,H2,IDZ,IN3,H3,D3,ICA,IPP,IPK)
      IF(IPK.EQ.1) GO TO 166
      M=M+1
      IF(M.GT.ILA) GO TO 166
      IF(TA.EQ.0) GO TO 35
      I=N
      IG=1
   37 IF(T(I).NE.0) GO TO 36
      I=I-1
      IG=IG+1
      IF(IG.EQ.N) GO TO 40
      IF(I.EQ.1) GO TO 20
      GO TO 37
   40 TA=0
      GO TO 35
   36 HN=HN1
      Z=1
   39 IF(IN1(Z).EQ.I) GO TO 38
      IN(Z)=IN1(Z)
      Z=Z+1
      GO TO 39
```

```
167 WRITE(3,104)
166 WRITE(3,340)
    DO 34 JP=1,JA
    WRITE(3,301) JP
    IZ=NN(JP)
    DO 350 I3=1,IZ
350 IN2(I3)=IN4(I3,JP)
    DO 41 I2=1,K
    H2(I2)=IHP(I2,JP)
    C1=H2(I2)
    C2=HOP(I2)
 41 DH(I2)=(C1-C2)/C2
 34 CALL DRUK(K,IZ,H2,IN2,DH)
    IF(IPP.EQ.1) GO TO 302
    WRITE(3,109) LZ
    GO TO 303
302 WRITE(3,106) LZ
303 CALL DRUK(K,NNM,H3,IN3,D3)
    STOP
100 FORMAT(8I10)
101 FORMAT(1H1,10X,42H***NETWORK MULTICRITERION OPTIMIZATION*** /
   1  /18X,16H******DATA******/
   2  40H NUMBER OF OBJECTIVE FUNCTIONS...........,I10/
   3  40H NUMBER OF   NODES.....................,I10/
   4  40H ASSUMED GREAT NUMBER.................,I10/
   5  40H NUMBER OF SUBOPTIMAL PATHS CONSIDERED...,I10/
   6  40H INTERMEDIATE OUTPUT...................,I10/
   7  40H ACCURACY OF COMPARING OF INCREMENTS.....,I10)
102 FORMAT(/12H MATRIX  BN(,I2,7H,I,J) =/)
103 FORMAT(/22H THE SHORTEST PATH FOR,I4,20H  OBJECTIVE FUNCTION)
104 FORMAT(56H NUMBER OF PARETO OPTIMAL PATHS HAS EXCEEDED ARRAY DIMEN
   1  5HSIONS)
105 FORMAT(1X,45(1H-)/
   1  46H PAYOFF TABLE FOR FUNCTION RELATIVE INCREMENTS /1X,45(1H-))
106 FORMAT(/33H THIS IS THE MIN-MAX OPTIMAL PATH/6H LZ = ,I5)
107 FORMAT(1X,8E15.5)
109 FORMAT(/37H THIS MAY BE THE MIN-MAX OPTIMAL PATH/6H LZ = ,I5)
110 FORMAT(1X,I3,3X,12I10/(7X,12I10))
111 FORMAT(17H ITERATION NUMBER ,I6)
114 FORMAT(18X,15H****RESULTS**** )
115 FORMAT(/51H SUBOPTIMAL PATHS FOR THE FIRST OBJECTIVE FUNCTION )
301 FORMAT(16H SOLUTION NUMBER,I6)
340 FORMAT(/21H PARETO OPTIMAL PATHS )
    END
C
C
C
C   SUBROUTINE MINMAX SELECTS THE MIN-MAX OPTIMAL PATH
C
    SUBROUTINE MINMAX(K,NNM,IZ,M,LZ,HOP,IN2,H2,IDZ,IN3,H3,D3,ICA,IPP,
   1IPK)
    INTEGER HOP(5),IN2(30),H2(5),IN3(30),H3(5)
    DIMENSION D3(5),DH(5),IDH(5),IB(5),ID(5),IDZ(5)
    IND=0
    I2=1
  1 C1=H2(I2)
    C2=HOP(I2)
    DH(I2)=(C1-C2)/C2
    IDH(I2)=DH(I2)*10.**ICA
    IF(IDH(I2).NE.0) GO TO 2
    IND=IND+1
  2 I2=I2+1
    IF(I2.LE.K) GO TO 1
    IF(IND.LT.K) GO TO 3
    DO 4 I4=1,K
    D3(I4)=0
```

```
    4 H3(I4)=H2(I4)
      DO 5 I3=1,IZ
    5 IN3(I3)=IN2(I3)
      NNM=IZ
      IPP=1
      IPK=1
      GO TO 6
    3 I3=0
      DO 7 I2=1,K
    7 IB(I2)=0
   11 I2=1
      IDA=0
    9 IF(IB(I2).NE.0) GO TO 8
      IF(IDA.GT.IDH(I2)) GO TO 8
      IDA=IDH(I2)
      IS=I2
    8 I2=I2+1
      IF(I2.LE.K) GO TO 9
      I3=I3+1
      ID(I3)=IDA
      IB(IS)=1
      IF(I3.GT.K) GO TO 10
      GO TO 11
   10 I3=1
   13 IF(ID(I3).LT.IDZ(I3)) GO TO 20
      I3=I3+1
      IF(I3.LE.K) GO TO 13
      GO TO 30
   20 I5=I3
   21 I5=I5-1
      IF(I5.EQ.0) GO TO 12
      IF(ID(I5).EQ.IDZ(I5)) GO TO 21
      GO TO 30
   12 DO 31 I3=1,IZ
   31 IN3(I3)=IN2(I3)
      LZ=M
      NNM=IZ
      DO 32 I4=1,K
      IDZ(I4)=ID(I4)
      H3(I4)=H2(I4)
   32 D3(I4)=DH(I4)
   30 IF(IDH(1).GT.IDZ(1))IPP=1
    6 RETURN
      END
C
C
C     SUBROUTINE PARETO SELECTS THE SET OF PARETO OPTIMAL PATHS
C
      SUBROUTINE PARETO(K,NN,IZ,JA,IN2,H2,IN4,IHP)
      INTEGER NN(100),IN2(30),H2(5),IN4(30,100),IHP(5,100)
      J=1
    4 KA=0
      DO 1 I=1,K
      IF(H2(I).LE.IHP(I,J)) KA=KA+1
    1 CONTINUE
      IF(KA.EQ.K) GO TO 2
      IF(KA.EQ.0) GO TO 3
      J=J+1
      IF(J.LE.JA) GO TO 4
      JA=JA+1
      NN(JA)=IZ
      DO 5 J1=1,IZ
    5 IN4(J1,JA)=IN2(J1)
      DO 6 I=1,K
    6 IHP(I,JA)=H2(I)
```

```
      GO TO 3
    2 NN(JA)=IZ
      DO 7 J1=1,IZ
    7 IN4(J1,J)=IN2(J1)
      DO 8 I=1,K
    8 IHP(I,J)=H2(I)
    3 RETURN
      END
C
C
C
C     SUBROUTINE DRUK PRINTS THE RESULTS OF CALCULATIONS

      SUBROUTINE DRUK(K,IZ,IPH,IPN,PD)
      DIMENSION IPH(5),IPN(30),PD(5),IPK(30)
      WRITE(3,1) (IPH(I),I=1,K)
      IF(PD(K+1).EQ.0.) GO TO 5
      WRITE(3,2) (PD(I),I=1,K)
    5 J=IZ+1
      DO 4 I=1,IZ
      J=J-1
    4 IPK(I)=IPN(J)
      WRITE(3,3) (IPK(I),I=1,IZ)
      RETURN
    1 FORMAT(34H VALUES OF THE OBJECTIVE FUNCTIONS/1X,10I12)
    2 FORMAT(43H VALUES OF THE FUNCTION RELATIVE INCREMENTS /1X,8E15.5)
    3 FORMAT(13H NODE NUMBERS/1X,20I6)
      END
```

<u>Output example</u>

NETWORK MULTICRITERION OPTIMIZATION

******DATA******

```
NUMBER OF OBJECTIVE FUNCTIONS...........          2
NUMBER OF   NODES.......................          8
ASSUMED GREAT NUMBER....................       1000
NUMBER OF SUBOPTIMAL PATHS CONSIDERED..         10
INTERMEDIATE OUTPUT.....................          0
ACCURACY OF COMPARING OF INCREMENTS....          5
```

MATRIX BN(1,I,J) =

1	20	25	24	20	1000	1000
2	1000	15	22	1000	1000	1000
3	10	20	1000	1000	1000	1000
4	1000	20	17	1000		
5	1	12	25			
6	1000	35				
7	16					

MATRIX BN(2,I,J) =

1	5	10	11	26	1000	1000
2	1000	4	20	1000	1000	1000
3	10	6	1000	1000	1000	1000
4	1000	5	9	1000		
5	1	7	12			
6	1000	8				
7	10					

****RESULTS****

```
THE SHORTEST PATH FOR    1  OBJECTIVE FUNCTION
VALUES OF THE OBJECTIVE FUNCTIONS
           45              38
NODE NUMBERS
     1     5     8

THE SHORTEST PATH FOR    2  OBJECTIVE FUNCTION
VALUES OF THE OBJECTIVE FUNCTIONS
           90              22
NODE NUMBERS
     1     2     4     6     8
-----------------------------------------------
PAYOFF TABLE FOR FUNCTION RELATIVE INCREMENTS
-----------------------------------------------
     0.              .10000E+01
     .72727E+00      0.

PARETO OPTIMAL PATHS
SOLUTION NUMBER        1
VALUES OF THE OBJECTIVE FUNCTIONS
           45              38
VALUES OF THE FUNCTION RELATIVE INCREMENTS
     0.              .72727E+00
NODE NUMBERS
     1     5     8
SOLUTION NUMBER        2
VALUES OF THE OBJECTIVE FUNCTIONS
           56              35
VALUES OF THE FUNCTION RELATIVE INCREMENTS
     .24444E+00      .59091E+00
NODE NUMBERS
     1     5     6     8
SOLUTION NUMBER        3
VALUES OF THE OBJECTIVE FUNCTIONS
           57              30
VALUES OF THE FUNCTION RELATIVE INCREMENTS
     .26667E+00      .36364E+00
NODE NUMBERS
     1     4     7     8
SOLUTION NUMBER        4
VALUES OF THE OBJECTIVE FUNCTIONS
           68              28
VALUES OF THE FUNCTION RELATIVE INCREMENTS
     .51111E+00      .27273E+00
NODE NUMBERS
     1     2     4     7     8
SOLUTION NUMBER        5
VALUES OF THE OBJECTIVE FUNCTIONS
           70              28
VALUES OF THE FUNCTION RELATIVE INCREMENTS
     .55556E+00      .27273E+00
NODE NUMBERS
     1     3     5     8

THIS IS THE MIN-MAX OPTIMAL PATH
LZ =     4
VALUES OF THE OBJECTIVE FUNCTIONS
           57              30
VALUES OF THE FUNCTION RELATIVE INCREMENTS
     .26667E+00      .36364E+00
NODE NUMBERS
     1     4     7     8
```

Symbols and notations

$\bigwedge\limits_{x}$ general quantifier 'for every x'.

$\bigvee\limits_{x}$ detail quantifier 'there is an x'.

$\wedge$ conjunction.

$\vee$ alternative.

$\in$ an element of.

ϕ empty set.

$\{\ \}$ set.

$\subset$ inclusion.

$\equiv$ is defined as.

$|\ |$ absolute value of a scalar.

$|$ given that.

Π product operator.

d_j the jth path in the graph G.

d^* optimal path.

D set of all paths in the graph G.

D^p set of Pareto optimal paths.

E^k k-dimensional space of objective functions.

E^n n-dimensional space of decision variables.

f function defined on the arcs of the graph G.

$\bar{f}(\bar{x})$ vector of objective functions.

$\bar{f}^0$ ideal vector of objective functions.

F map of the set X in E^k.

F^p map of the set X^p in E^k.

F^t t-directional shadow for the range F.

$g(\bar{x})$ inequality constraints (distinguished by subscripts).

G symbol of graph $G = \langle U, R \rangle$.

$h(\bar{x})$ equality constraints (distinguished by subscripts).

I set of indices for all objective functions.

U set of graph nodes.

R double argument relation defined on the set U.

S symbol of network $S = \langle U, R, f \rangle$.

$\bar{x}$ vector of decision variables.

$\bar{x}^{0(i)}$ vector of decision variables which minimizes the ith objective function.

$\bar{x}^*$ optimal vector of decision variables.

X set of feasible solutions.

X^p set of Pareto optimal solutions.

$\bar{z}(\bar{x})$ vector of relative function increments.

Author index

Subject index